Contents

How to use this book

Each page has a title telling you what it is about.

Instructions look like this. Always read these carefully before starting.

This shows you how to set out your work. The first question is done for you.

Sometimes you need materials to help you.

Ask your teacher if you need to do these.

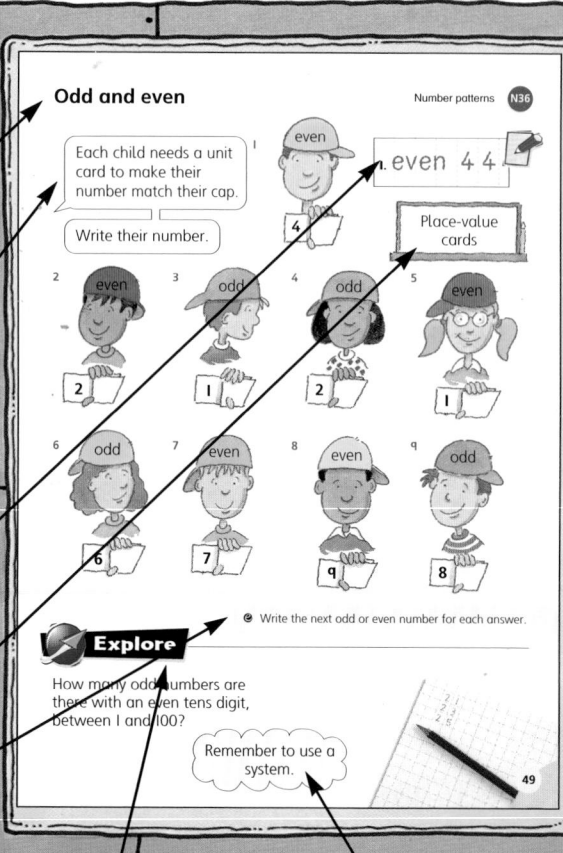

This shows that the activity is an **Explore**. Work with a friend.

Sometimes there is a **Hint** to help you.

This means you must decide how to set out your work and show your workings.

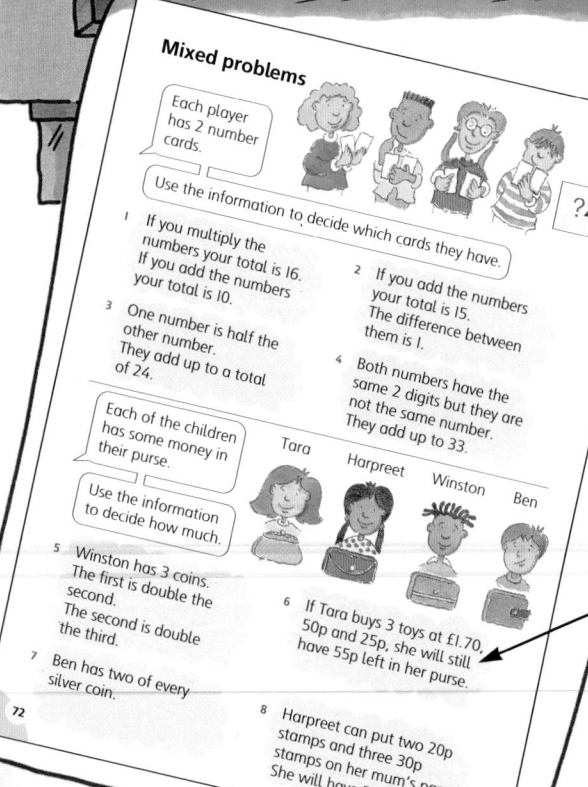

Read the word problems very carefully.
Decide how you will work out the answers.

Multiples of 10

The winning tickets in the draw are all multiples of ten.

Write the winning numbers.

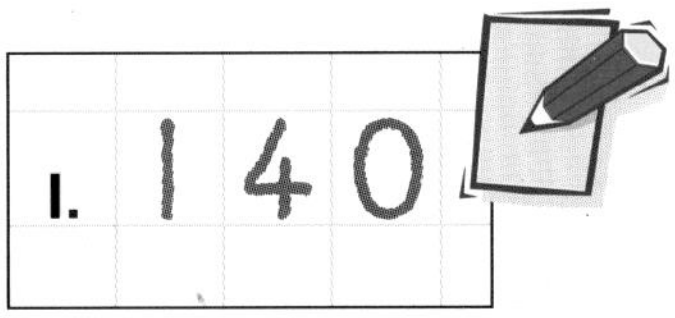

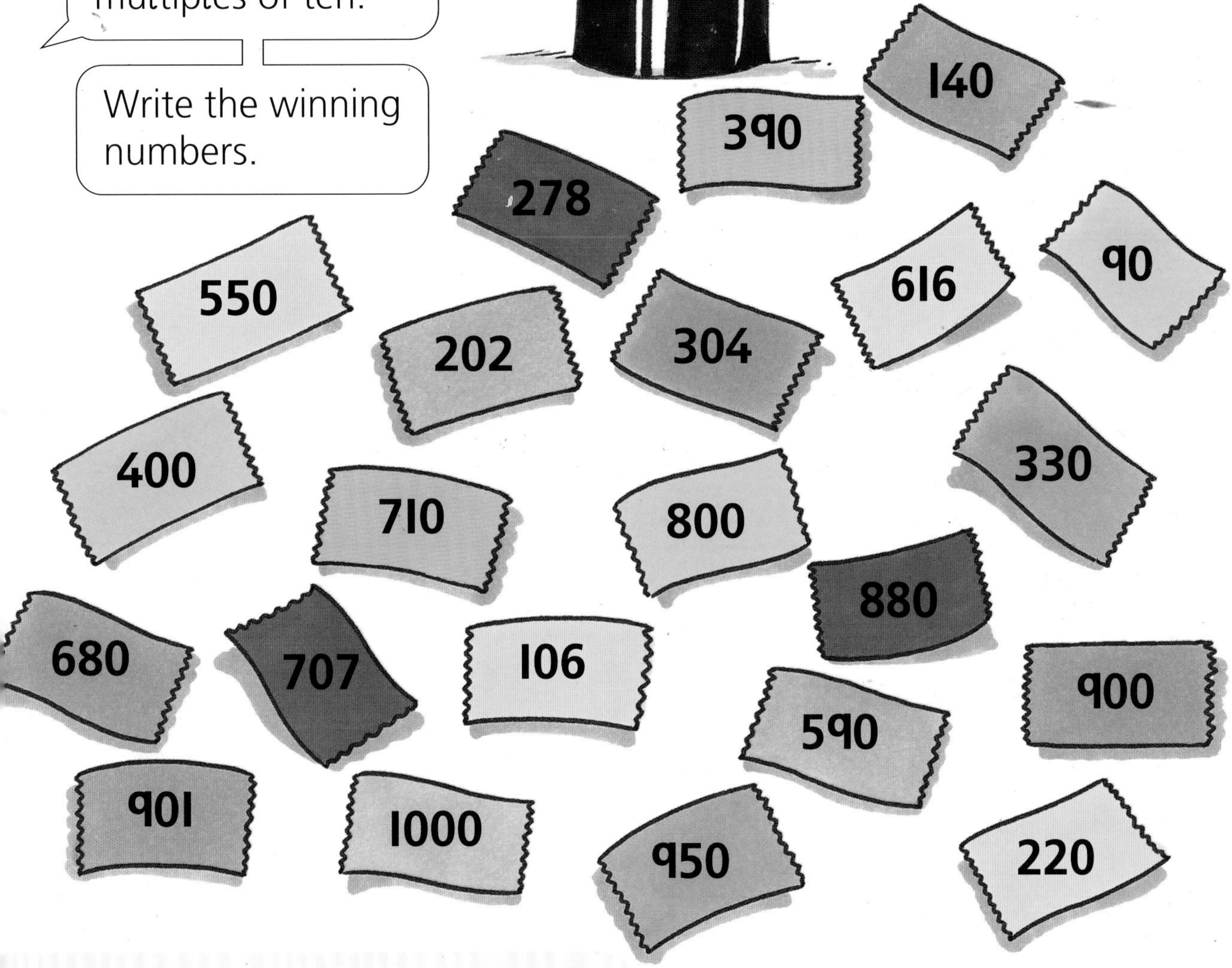

I. 140

390

278

140

550

616

90

202

304

400

710

800

330

680

707

106

880

900

590

901

1000

950

220

ⅇ Write the tickets that are multiples of 100.

Explore

Write all the multiples of 10 between 218 and 456. How many are there?

How many of these numbers are multiples of 100?

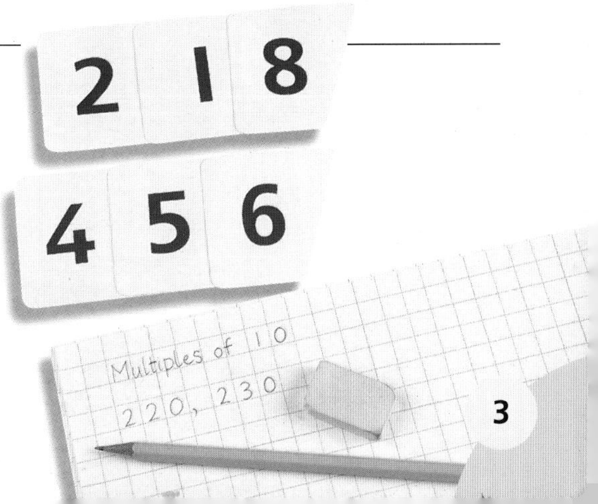

2 1 8

4 5 6

Multiples of 10
220, 230

3

Counting in 50s

Write the number where each fly has landed.

1a. 5 5 0

1

a

| | | | | | | |
400   500   600   700

2

b   c   d

700   800   900   1000   1100

3

e   f

900   1000   1100   1200   1300

4

g   h   i

1000   1100   1200   1300   1400

5

j   k

1200   1300   1400   1500   1600

Count on in 100s.
Write the next 3 numbers.

6 464

6. 5 6 4
   6 6 4
   7 6 4

7 347

8 579

9 545

10 494

11 688

12 299

13 342

14 656

15 707

16 519

17 202

18 98

🄔 Write the numbers 100 less.

4

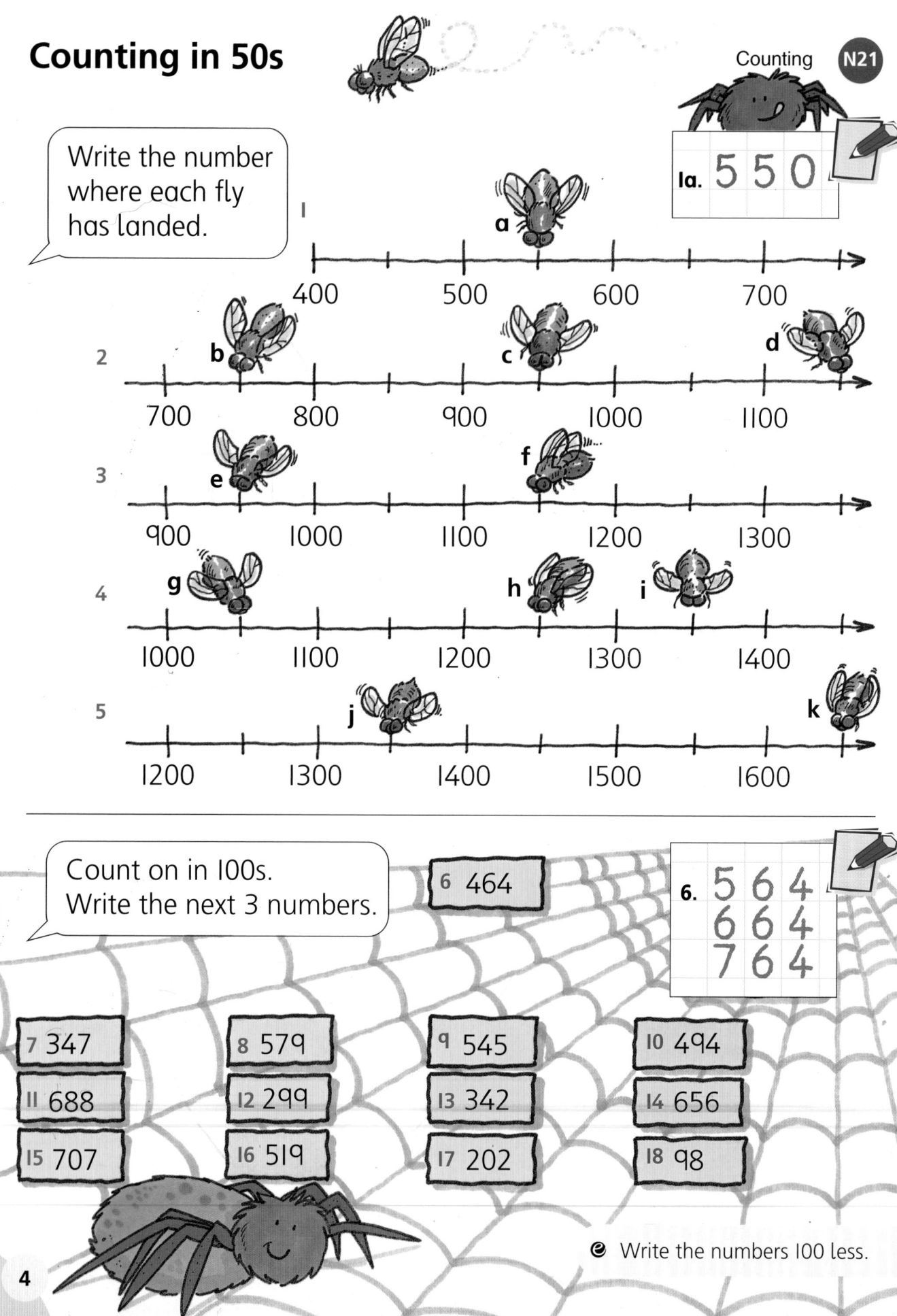

Counting in 50s

850, 900, 950, 1000

Count in 50s.
Write the next
4 numbers.

1 **750** **800**

2 **600** **650**

3 **250** **300**

4 **50** **100**

5 **300** **350**

6 **950** **900**

7 **700** **650**

Problems

8

Sara has **£2.50** pocket money.
Dad gives her **50p** more.
How much now?

9 The kitten
weighs **350 g**.

She puts on **50 g**
each week.

How heavy is she
after **3 weeks**?

10

Andrew's snake was **25 cm** long.
It grew **50 cm**.
How long is it now?

11

The balloon is **750 m**
above the ground. It
falls **50 m** each
minute.

How high is it
after **5 minutes**?

Odd and even

Count the eggs in each set.
Write odd or even.

1   1. 7 → odd

2   3   4

5   6   7

Copy and write the next 5 even numbers.

8. 30, 32, 34, 36 …

8   30   32

9   6   8

10   24   26

11   38   40

Write the next 5 odd numbers.

12   3   5   ?

13   21   23

14   37   39

Odd and even

Write odd or even for each bus number.

1. $42 \rightarrow$ even

2  16

3  43

4  21

5  34

6  49

7  33

8  51

9  11

10  10

11  14

12  32

13  50

14  25

15  17

16  38

17  47

 **Explore**

Use the number cards shown.

Make different 2-digit numbers.

How many are odd? How many are even?

Repeat with 3 different cards.

Can you see any patterns?

Odd and even

Write the even numbers up to 50, in a chain.

2 4 6 8 10 12

e Write the odd numbers from 49 to 1.

1

Carol has lots of pencils.

There are more than **45** but less than **50**.

It is an even number.

How many could she have?

2

Sam has a large family.

There are an even number of people.

The digits of the number add up to **1**.

How many in Sam's family?

Problems

3

Sumi's age is an odd number.

Sumi's brother is **5 years** older than her. His age is an even number.

He is younger than **20**.

How old could Sumi be?

4

Bob's team played **2** matches.

The total goals scored was an odd number.

It was less than half of **12** and more than **4**.

How many goals were scored?

Pairs to 100

Write pairs of balloons that total 100.

1. 3 5 and 6 5

40

35
25
15

45

65

30

55

20
85
60
75

70
80

95
5

Copy and complete.

9. 7 0 + 3 0 = 1 0 0

9 70 +  = 100

10 75 +  = 100

11  + 35 = 100

12 45 +  = 100

13 100 − 5 =

14 100 − 85 =

9

Pairs to 100

Megan has a £1 coin. How much change from buying each key ring?

1 **30p**

I. $£1.00 - 30p = 70p$

2 **75p**

3 **80p**

4 **45p**

5 **60p**

6 **55p**

7 **85p**

8 **90p**

9 **65p**

10 **25p**

ℯ How much change from a £2 coin?

Explore

Use 5p and 10p coins.

Find the different ways to make £1.

Remember to use a system.

Adding 2-digit numbers

Copy and complete.

1. 60 + 40 =

1. $60 + 40 = 100$

2. 59 + 40 =

3. 35 + 65 =

4. 71 + 32 =

5. 75 + 26 =

6. 85 + 14 =

7. 79 + 12 =

8. 49 + 49 =

9. 75 + 35 =

10. 15 + 87 =

11. 29 + 69 =

12. 61 + 42 =

13. 19 + 90 =

Problems

Disco ticket £1.00

14. Bella has **£1.25**.

Her brother George has **65p**.

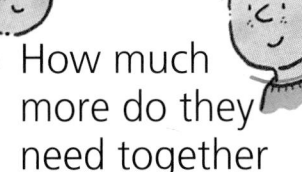

How much more do they need together for a ticket each?

15. Wesley has **55p**.

Ben has **45p**.

How much do they need together to buy a ticket each?

16. Asha has **75p**

How much more does she need?

17. Kris has **95p**.

Jo has **95p**.

How much more do they need between them?

11

Doubling

Double each number.

1 | **2 3**

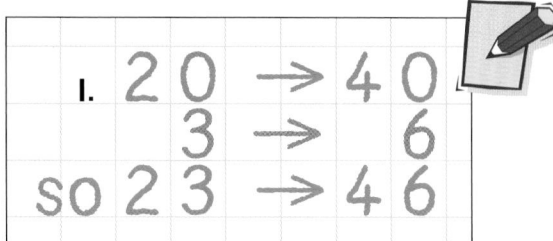

Double the tens, then double the units ... add them together.

2 | **2 4**

3 | **3 1**

4 | **4 2**

5 | **3 2**

6 | **1 2**

7 | **2 1**

8 | **3 4**

9 | **5 2**

10 | **4 3**

Double the amount.

Double it again.

11

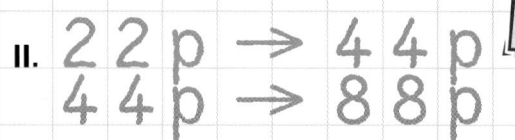

12

13

14

15

16

17

Doubling

How much for 2 of each pen?

1. **95p**

$$95p \rightarrow 190p$$
$$190p \rightarrow £1.90$$

2. **35p**

3. **45p**

4. **55p**

5. **5p**

6. **15p**

7. **75p**

8. **25p**

9. **65p**

10. **85p**

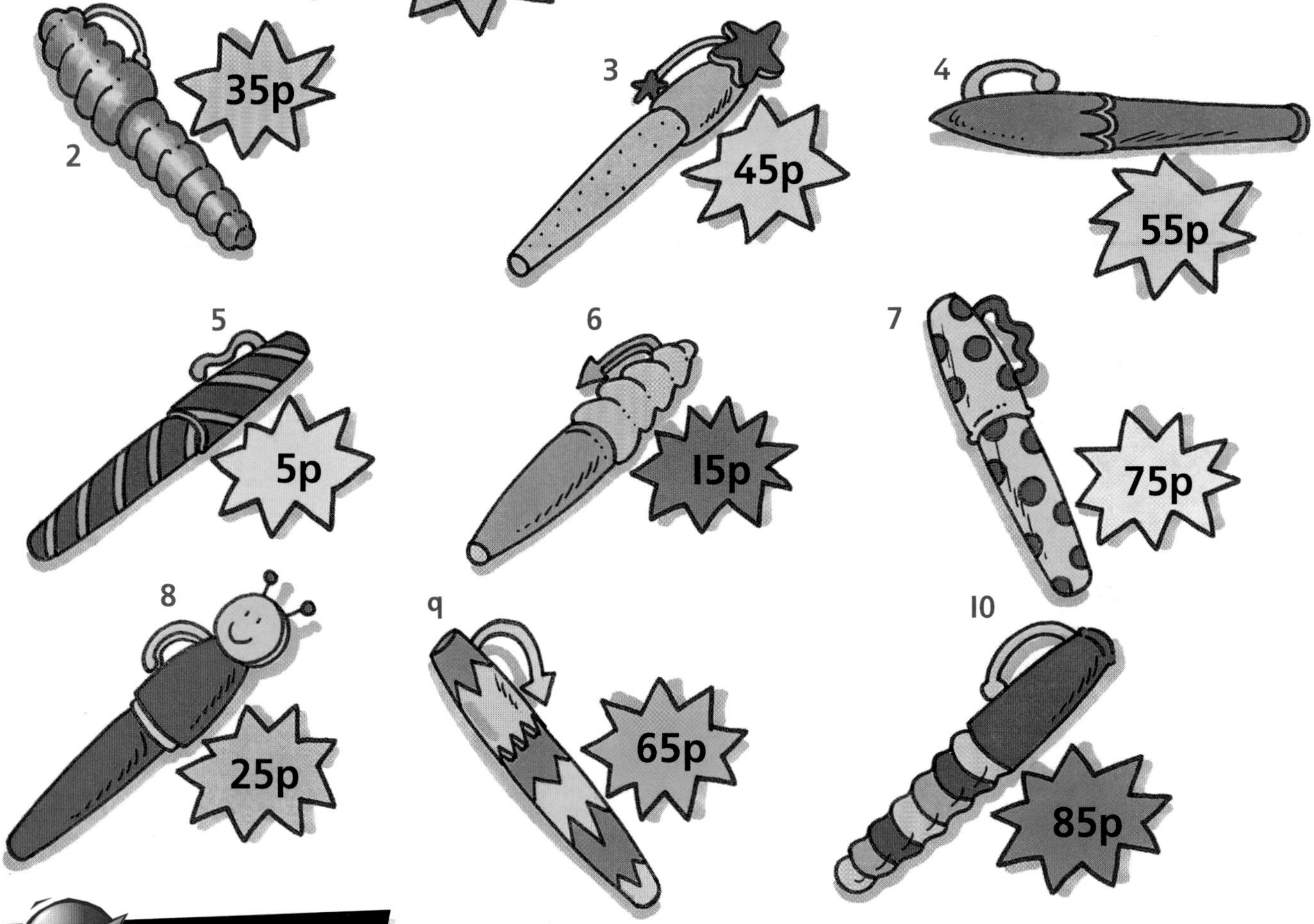

Explore

Use all the hundreds place-value cards
and the ⬚ 5 ⬚ units card.

Choose a hundreds card. Place the 5 on top.

You have a 3-digit multiple of 5.

Double it and write the answer.

Use each hundred card in turn.

Is there a pattern?

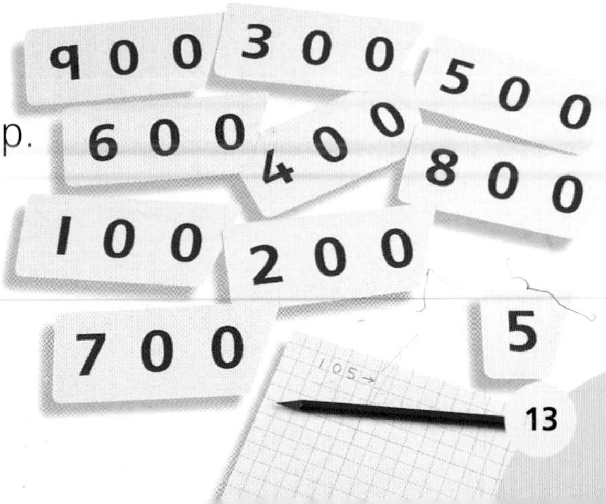

900 300 500
600 400 800
100 200
700 5

13

Doubling

Write the answers.

1. double 25 ⟶
   double 26 ⟶

1. $25 \rightarrow 50$
   $26 \rightarrow 52$

2. 15 doubled is
   16 doubled is

3. 35 + 35 =
   36 + 36 =

4. double 12 is
   double 13 is

5. 45 doubled is
   46 doubled is

6. 25 + 25 =
   25 + 26 =

7. 24 doubled is
   23 doubled is

8. double 50 is
   double 52 is

9. 50 + 50 =
   49 + 49 =

10. 30 + 30 =
    29 + 30 =

11. double 100 =
    double 99 =

Problems

12.
    Tom has **43p.**
    His mum doubles it.
    How much has he now?

13.
    Josh has **34p.**
    His step-dad doubles it.
    How much has he now?

14.
    Samira has **102p.**
    Her gran doubles it.
    How much has she now?

Fives

Write the 5s to 100.

5   10

Write how many fingers in each set of gloves.

1

1. 3 × 5 = 15

2

3

4

5

6

7

8

9

Copy and complete.

10   1 × 5 =

11   4 × 5 =

10. 1 × 5 = 5

12   5 × 5 =

13   9 × 5 =

14   8 × 5 =

15   3 × 5 =

16   6 × 5 =

17   10 × 5 =

18   7 × 5 =

19   2 × 5 =

@ Write these facts in order.

15

Tens

Copy and complete.

1. $1 \times 10 =$

1. $1 \times 10 = 10$

2. $6 \times 10 =$

3. $4 \times 10 =$

4. $9 \times 10 =$

5. $3 \times 10 =$

6. $7 \times 10 =$

7. $5 \times 10 =$

8. $10 \times 10 =$

9. $8 \times 10 =$

10. $2 \times 10 =$

ℰ Write these facts in order.

Win 10p for each skittle knocked down.

11. $6 \times 10p = 60p$

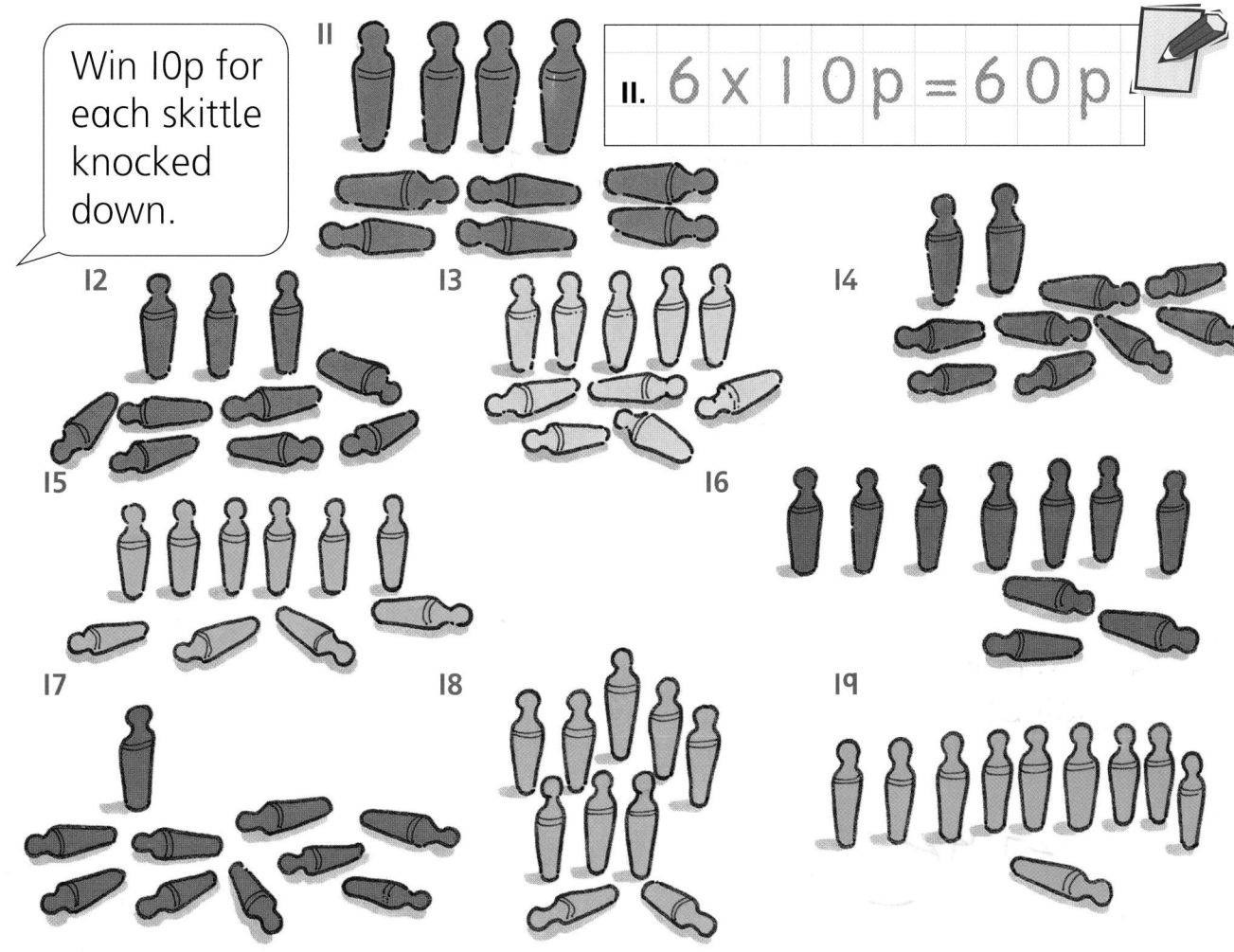

ℰ Write the winnings if each skittle wins 5p.

Fives and tens

> Write how many 5p coins you need to buy each stamp.

I 30p

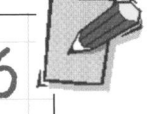

I. $30p \div 5p = 6$

2  15p

3  5p

4  40p

5  25p

6  45p

7 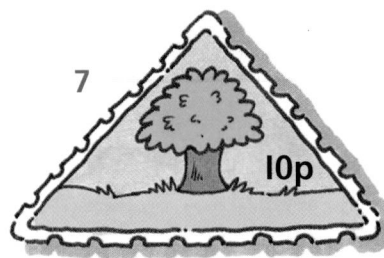 10p

8  20p

q  50p

10  35p

Problems

11

Zamara saves **5p** each day.

How many days before she can buy the notebook?

45p

12

Mick helps his dad.

He has **20** cans to pack.

5 cans go in each box.

How many boxes?

Explore

Gita has in her purse the coins shown.
What amounts can she pay exactly?

Multiplying

Write how many stamps in each set.

I. five twos
5 x 2 = 10

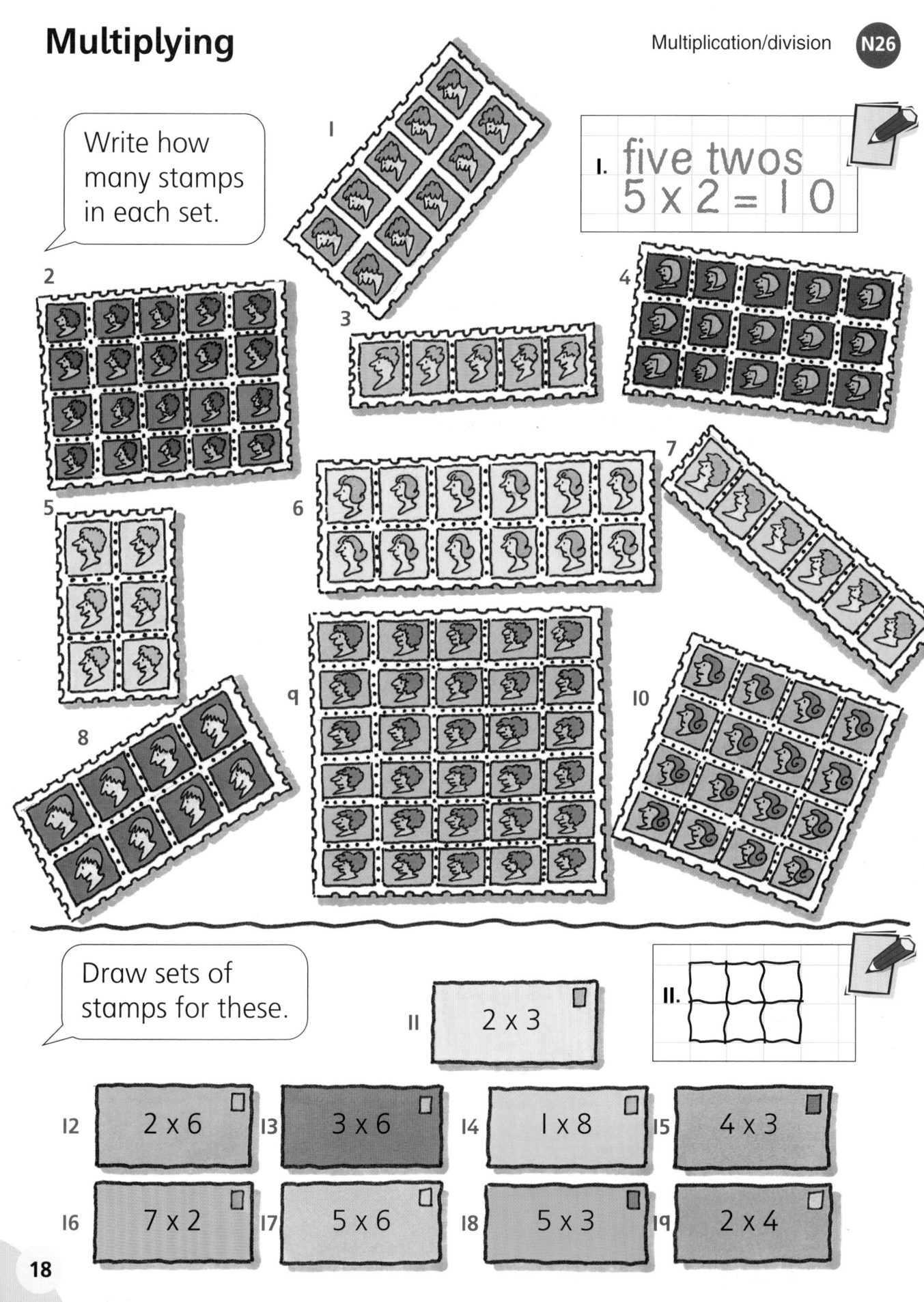

Draw sets of stamps for these.

II.

11. 2 x 3

12. 2 x 6
13. 3 x 6
14. 1 x 8
15. 4 x 3

16. 7 x 2
17. 5 x 6
18. 5 x 3
19. 2 x 4

Multiplying

1
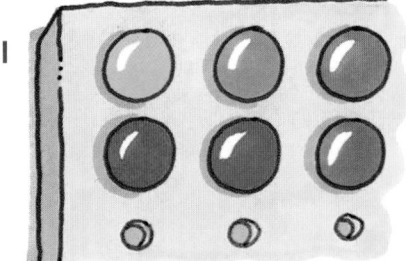

Write how many pegs are on the board.

1. $2 \times 3 = 6$
   $3 \times 2 = 6$

2
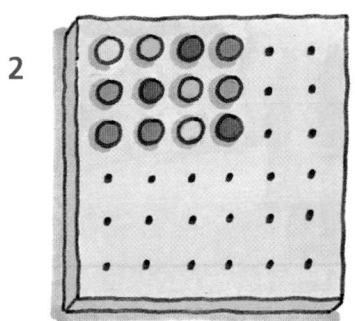

3
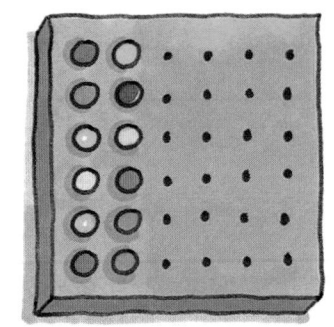

4
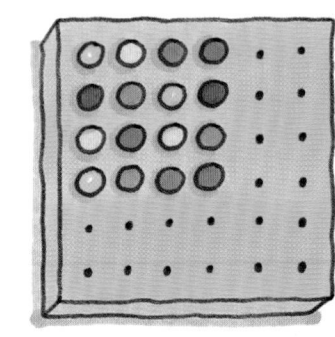

5
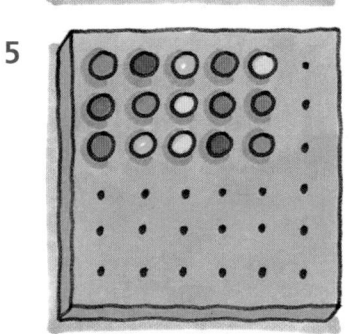

6
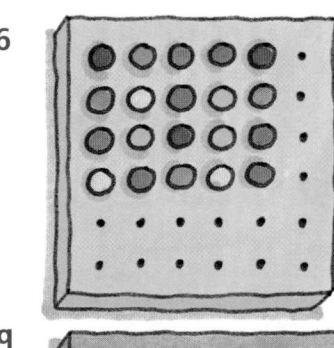

7
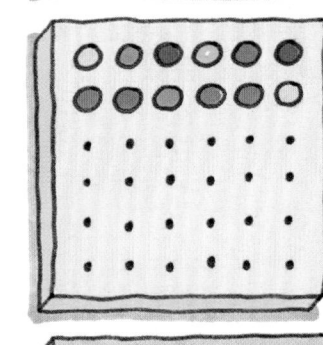

8
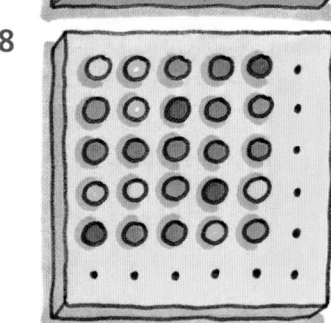

9
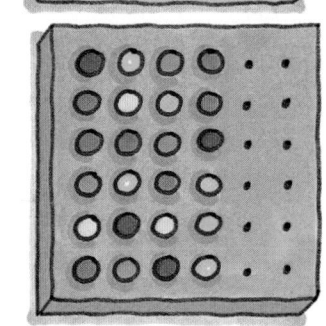

10
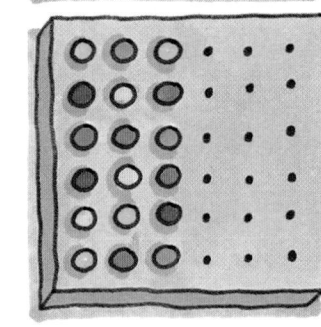

@ Make some pegboard multiplications of your own.

 **Explore**

Use 24 counters.

Place them all on squared paper to make a rectangle. Remove them and make a different rectangle.

How many different rectangles can you make?

Write a multiplication for each.

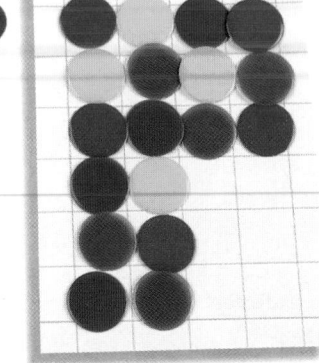

19

Multiplying and dividing

Draw fish to match the divisions.

Write a matching multiplication.

1 8 ÷ 4

2 10 ÷ 5

1.

$8 \div 4 = 2$
$2 \times 4 = 8$

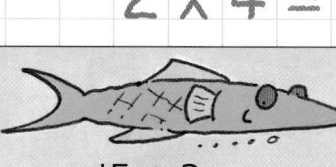

3 12 ÷ 3

4 6 ÷ 2

5 15 ÷ 3

6 20 ÷ 5

7 12 ÷ 6

8 20 ÷ 10

Copy and complete the multiplication table.

Use counters to help you.

$2 \times 4 = 8$

×	1	2	3	4	5	6
1						
2				8		
3						
4						
5						
6						

Threes

Copy and complete.

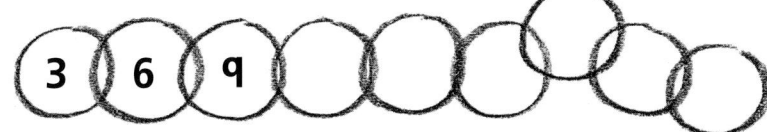

Write how many legs in each set of monsters.

1. $4 \times 3 = 12$

Copy and complete.

10 $1 \times 3 =$

10. $1 \times 3 = 3$

11 $5 \times 3 =$
12 $9 \times 3 =$
13 $6 \times 3 =$
14 $3 \times 3 =$
15 $2 \times 3 =$
16 $10 \times 3 =$
17 $8 \times 3 =$
18 $4 \times 3 =$
19 $7 \times 3 =$

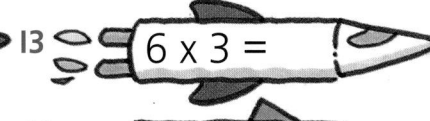
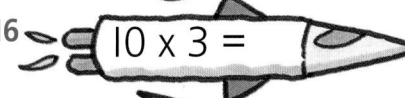
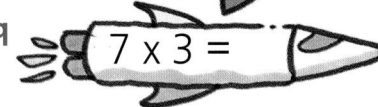

e Write these facts in order.

Threes

Each blast is worth 3 points. Write the total.

1

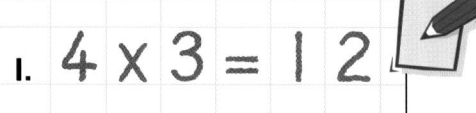
I. $4 \times 3 = 12$

2

3

4

5

6

7

Copy and complete.

8  $\times 3 = 6$

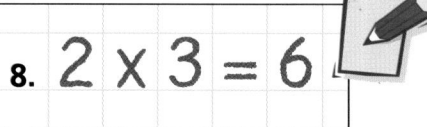
8. $2 \times 3 = 6$

9 $\times 3 = 18$

10 $\times 3 = 15$

11 $\times 3 = 9$

12 $\div 3 = 4$

13 $\div 3 = 10$

14 $\div 3 = 1$

15 $\times 3 = 27$

16 $\div 3 = 8$

17 $\times 3 = 21$

Threes

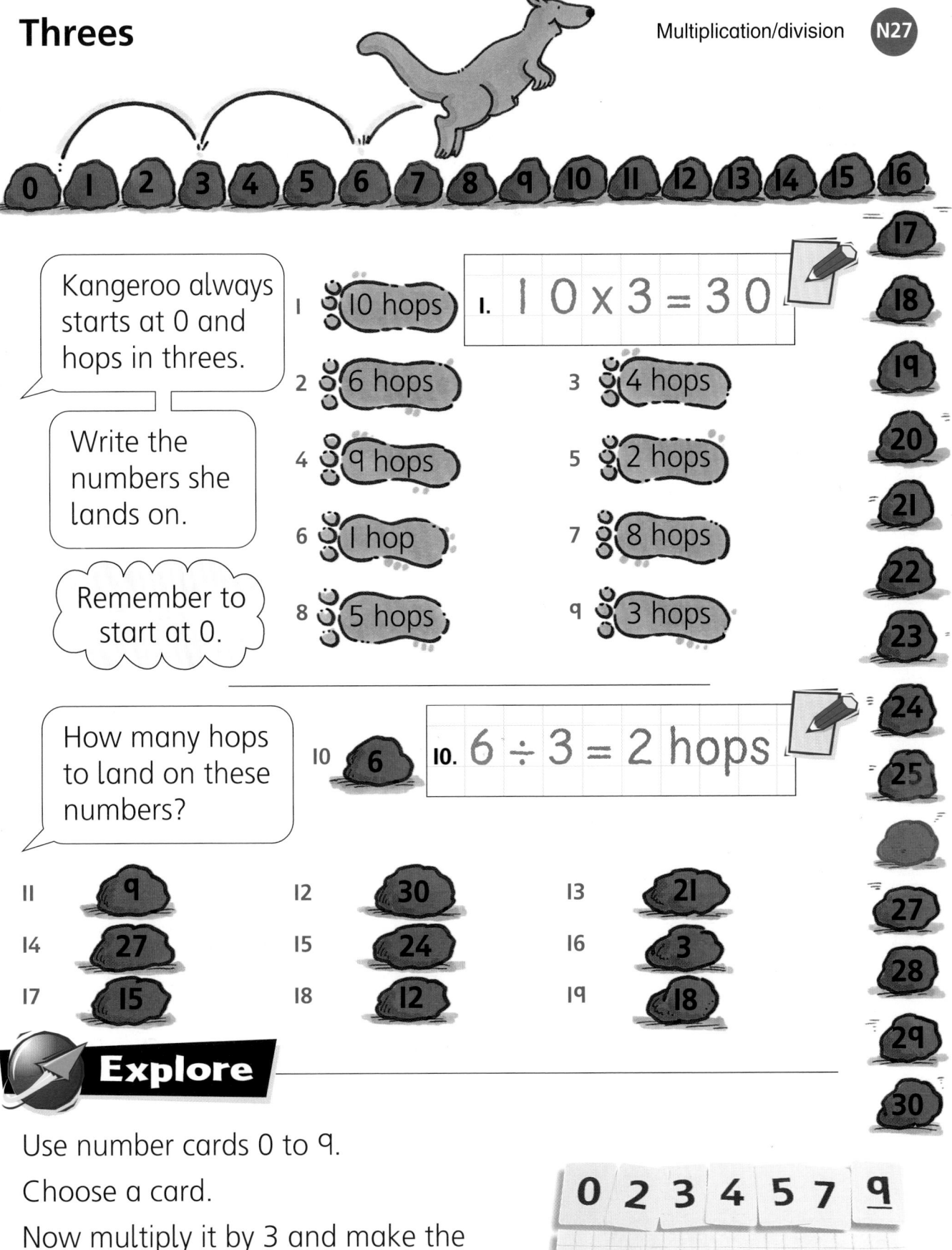

Kangeroo always starts at 0 and hops in threes.

Write the numbers she lands on.

Remember to start at 0.

1. 10 hops   1. $10 \times 3 = 30$

2. 6 hops   3. 4 hops

4. 9 hops   5. 2 hops

6. 1 hop   7. 8 hops

8. 5 hops   9. 3 hops

How many hops to land on these numbers?

10. 6   10. $6 \div 3 = 2$ hops

11. 9   12. 30   13. 21

14. 27   15. 24   16. 3

17. 15   18. 12   19. 18

Explore

Use number cards 0 to 9.

Choose a card.

Now multiply it by 3 and make the answer with the other cards.

How many different ways can you find?

0 2 3 4 5 7 9

6 → 1 8

23

Fractions

Write the fraction of each shape that is blue.

1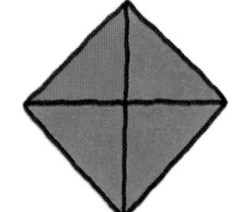

I. $\dfrac{3}{4}$

2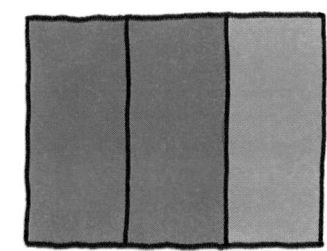

3

4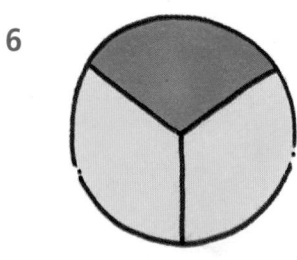

5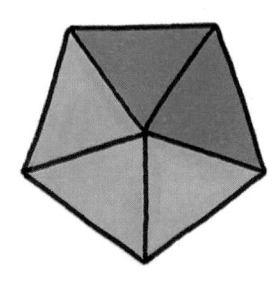

6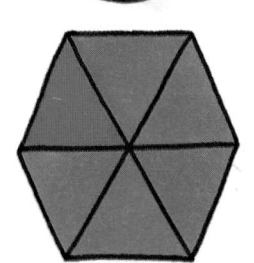

7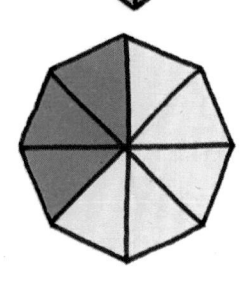

8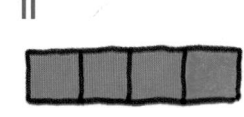

9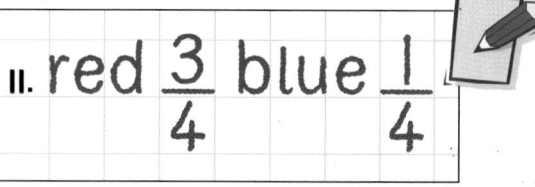

10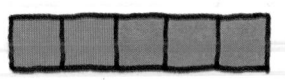

Write the fractions of each strip that are red and blue.

11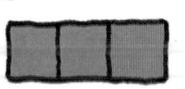

II. red $\dfrac{3}{4}$ blue $\dfrac{1}{4}$

12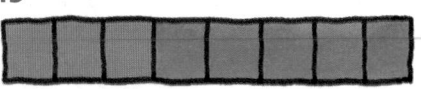

13

14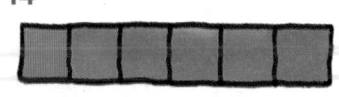

15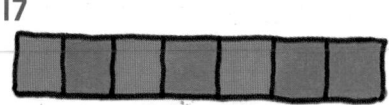

16

17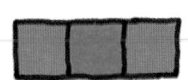

Fractions

Copy each and colour the fraction shown.

1 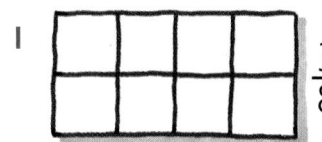 $\frac{3}{8}$

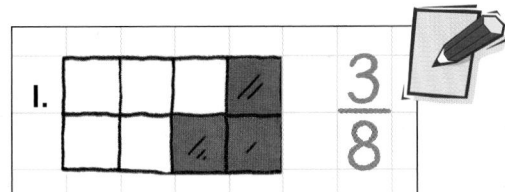
1. $\frac{3}{8}$

2 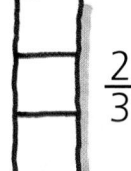 $\frac{2}{3}$

3 $\frac{3}{4}$

4 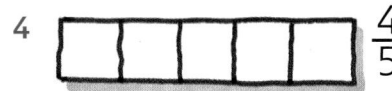 $\frac{4}{5}$

5 $\frac{3}{6}$

6 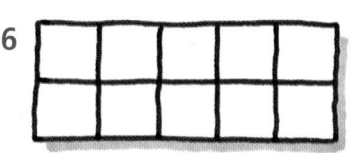 $\frac{4}{10}$

7  $\frac{6}{9}$

8 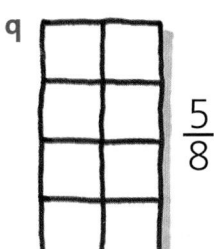 $\frac{4}{6}$

9 $\frac{5}{8}$

10 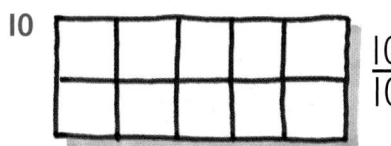 $\frac{10}{10}$

e Write the fraction of each grid not coloured.

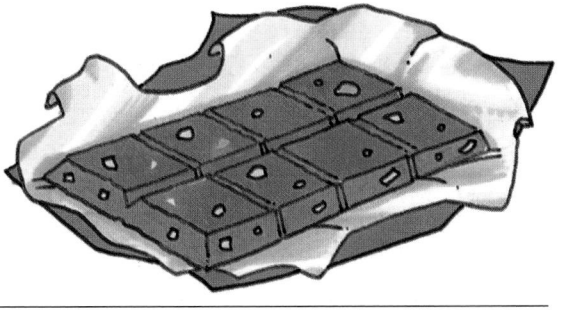

Explore

Use squared paper.

Draw different rectangles.

Colour the squares using 2 colours.

Write the fraction for each colour.

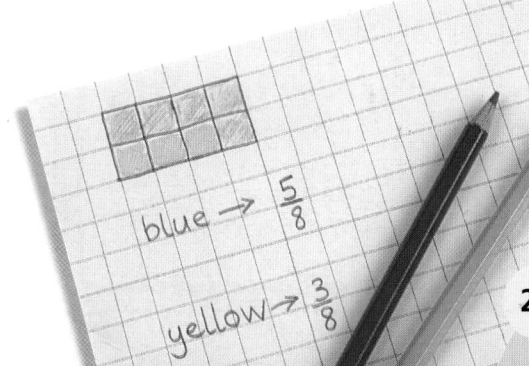

blue → $\frac{5}{8}$

yellow → $\frac{3}{8}$

25

Fractions

How many chocolate drops in the fraction shown?

1 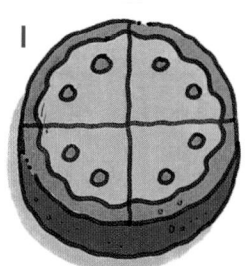 $\frac{1}{4}$

1. $\frac{1}{4}$ of $8 = 2$

2 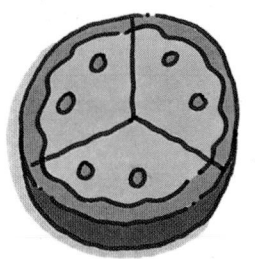 $\frac{1}{3}$

3 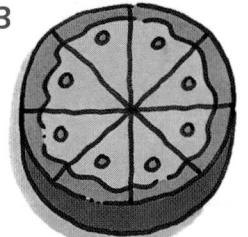 $\frac{3}{8}$

4 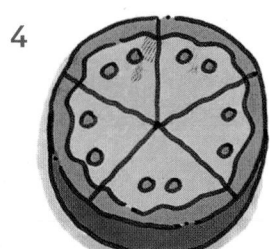 $\frac{2}{5}$

5  $\frac{5}{6}$

6 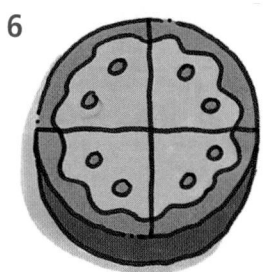 $\frac{3}{4}$

7 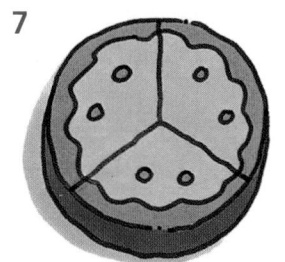 $\frac{2}{3}$

8 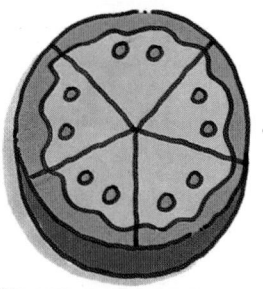 $\frac{4}{5}$

9  $\frac{1}{4}$

10 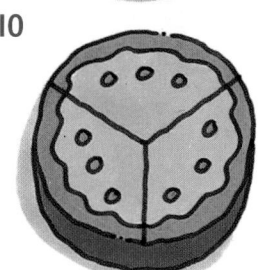 $\frac{2}{3}$

Problems

11 **15** children are playing catch.

$\frac{2}{3}$ are girls.

How many are boys?

12

Jane has **20** raisins.

She eats $\frac{3}{4}$ of them.

How many are left?

13

$\frac{2}{5}$ of the pizza has been eaten.

What fraction is left?

Nearest ten

Write the position of each flower.

1a. | 2 | 3 |

1

a   b   c

20   25   30

2

d   e   f   g   h

40   50   60

3

i   j   k   l   m

60   70   80   90   100

Write how much in each pile.

Round each answer to the nearest 10p.

4. | 16p → 20p |

4
5
6
7
8
9

Nearest hundred

Write the position of each bird.

1a. **1 3 0**

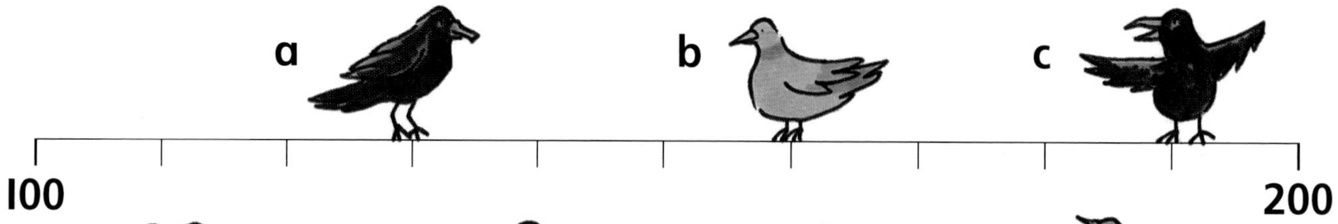

a   **b**   **c**

1
|
100   200

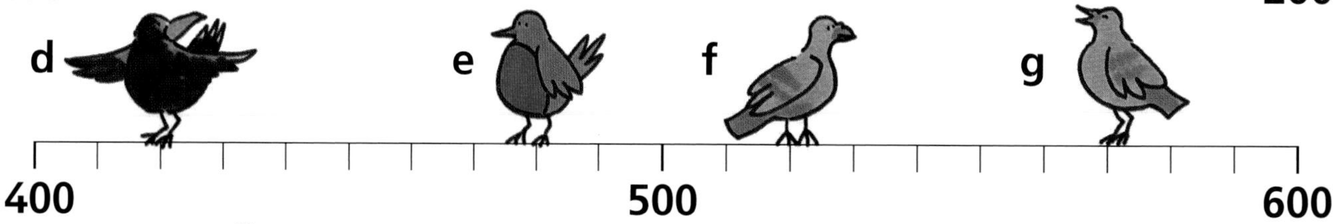

d   **e**   **f**   **g**

2
|
400   500   600

h **i**   **j**   **k**

3
|
200   300   400   500   600

e Round each answer to its nearest 100.

Round each weight to the nearest 100 grams.

4. **4 6 0 g → 5 0 0 g**

4 
460 g

5 
320 g

6
940 g

7
680 g

8
170 g

9
550 g

10
770 g

Nearest hundred

These are distances from London in miles.

Write each one to the nearest 100 miles.

I. 4 0 3 → 4 0 0 miles

1 Glasgow 403

171 Exeter   2

3 Derby 128

188 Hull   4

5 Newcastle 278

390 Ayr   6

7 Holyhead 264

84 Salisbury   8

9 Thurso 636

Explore

Jack has forgotten the number on his lock.
He knows it is three of these numbers.

3   7   4   9

Write all the possible 3-digit numbers it could be.

29

Adding multiples of 5

Write pairs of numbers to make 100.

Use multiples of 5.

Start with
5 + 🌸 = 100

5 + 95 = 100
10 +
15 +
20 +
25 +

Choose 2 cakes.

Write the cost.

Repeat 10 times, choosing different cakes.

1. $25p + 15p = 40p$

Cake Sale.

25p each

35p each

45p

15p each

5p each

55p

Adding multiples of 5

> Write the height of each flower at the end of May.

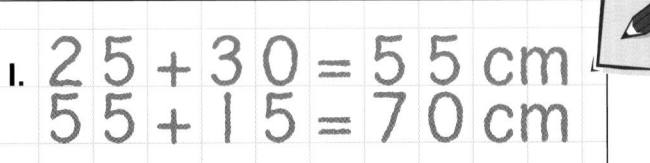

I. $25 + 30 = 55$ cm
$55 + 15 = 70$ cm

1

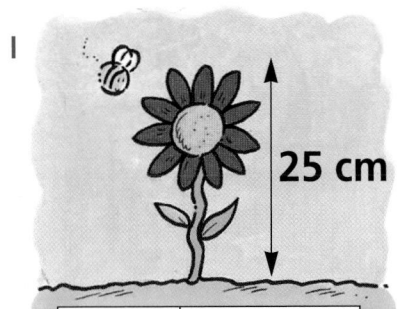

25 cm

	growth
April	30 cm
May	15 cm

2

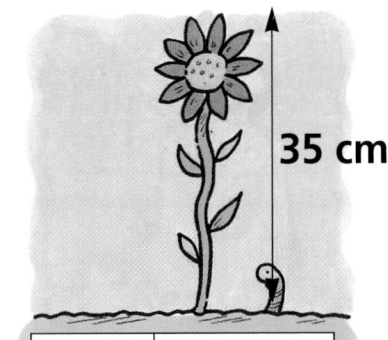

35 cm

	growth
April	50 cm
May	15 cm

3

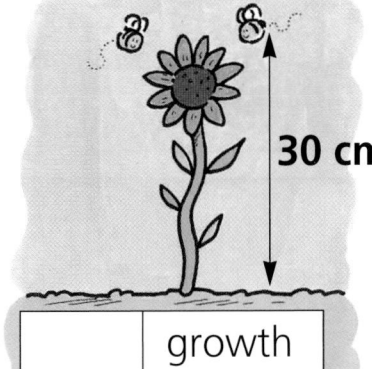

30 cm

	growth
April	45 cm
May	25 cm

4

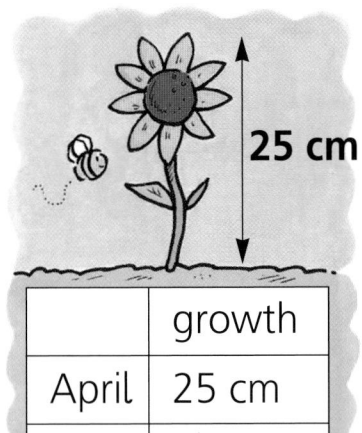

25 cm

	growth
April	25 cm
May	15 cm

5

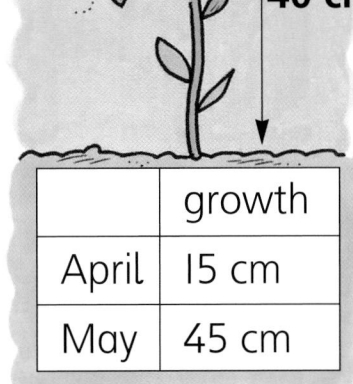

40 cm

	growth
April	15 cm
May	45 cm

6

15 cm

	growth
April	15 cm
May	15 cm

Explore

Use the three spinners shown.

Write the highest possible score.

Write the lowest possible score.

How many different scores are there between 50 and 75?

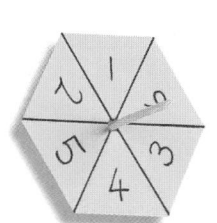

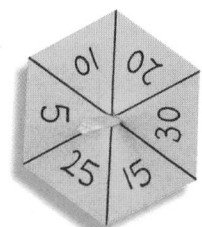

Each child throws 3 balls.

Write their scores.

Who wins?

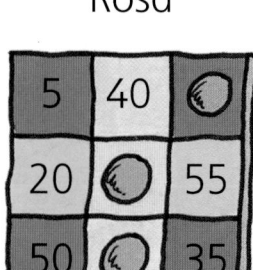

Jon

Ashley

🔘🔘	40	15
20	45	55
50	12	🔘

Ben

5	🔘	15
🔘	45	🔘
50	12	35

Rosa

5	40	🔘
20	🔘	55
50	🔘	35

Ana

5	🔘	15
🔘	🔘	55
50	12	35

Sumi

5	40	🔘
20	45	🔘
🔘	12	35

Jenny

5	40	15
20	🔘	🔘
50	12	🔘

Tom

5	40	15
20	🔘	55
🔘	12	🔘

Pete

5	40	🔘
20	🔘	55
🔘	12	35

Aziz

🔘	🔘	🔘
20	45	55
50	12	35

Bec

5	🔘	🔘
20	45	55
🔘	12	35

Jodi

🔘	40	🔘
20	45	55
🔘	12	35

Nick

🔘	40	🔘
20	45	55
50	🔘	35

Taking away 10, 20, 30

Write each clock time 20 minutes ago.

I. 3:28 → 3:08

1. 3:28

2. 6:46

3. 4:38

4. 8:57

5. 5:24

6. 10:31

7. 7:40

8. 3:29

9. 11:49

10. 2:48

11. 1:55

12. 3:46

13. 12:25

Copy and complete the table.

in	4	2	5	7	7	3		
out	1	2						

take away 30

in 42

out 12

in	42	57	73	95	39	64	87	66
out	12							

42

12

Taking away

Sale 25p off

Each toy car has 25p off the price.

Write the new prices.

1. $65p - 25p = 40p$

1  **65p**

2  **35p**

3  **75p**

4  **55p**

5  **85p**

6  **45p**

7  **95p**

8  **30p**

9  **50p**

10  **70p**

Problems

11

Sally had **45p**.
She has lost **15p**.
How much is left?

12

Sam has **60p**.
He buys a comic for **35p**.

How much is left?

13
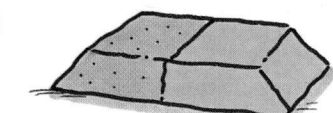
Jo has **65p**.
She buys a pencil for **40p** and a rubber for **5p**.
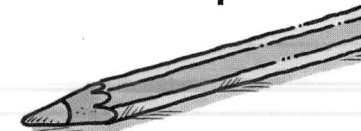
How much is left?

Taking away

Take away from each pile.	Use real coins.	I. $56p - 23p = 33p$ 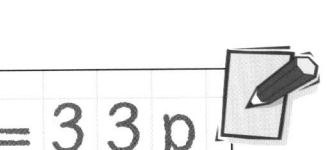

1 take away 23p

10p

1p

2 take away 31p

10p

1p

3 take away 24p

10p

1p

4 take away 15p

10p 1p

5 take away 41p

10p

1p

6 take away 65p

10p

1p

7 take away 42p

10p

1p

8 take away 28p

10p 1p

9 take away 13p

10p

1p

℮ Make up some money subtractions of your own.

Take each star number away from each circle number.

Write your answers in order, smallest to largest.

10. $75 - 14 = 61$

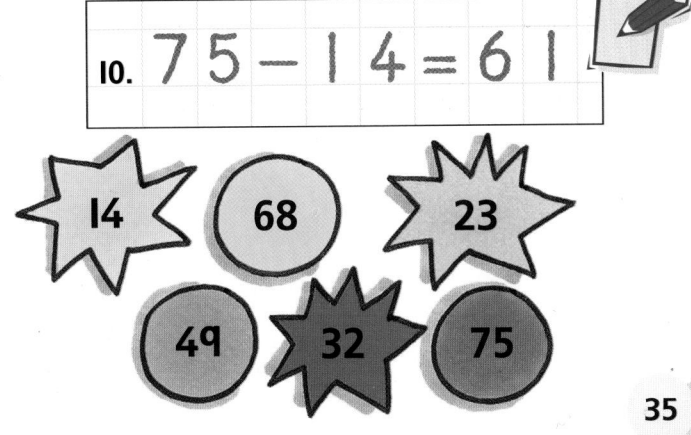

Difference

Find pairs of numbers the same colour.

I. $8 + 5 = 13$

Count from the smaller to the larger.

Write an addition.

1	2	3	4	5	6	7	8	9	10
11	12	13	14	15	16	17	18	19	20
21	22	23	24	25	26	27	28	29	30
31	32	33	34	35	36	37	38	39	40
41	42	43	44	45	46	47	48	49	50
51	52	53	54	55	56	57	58	59	60
61	62	63	64	65	66	67	68	69	70
71	72	73	74	75	76	77	78	79	80
81	82	83	84	85	86	87	88	89	90
91	92	93	94	95	96	97	98	99	100

Count from one flag to the next.

Write an addition and a subtraction.

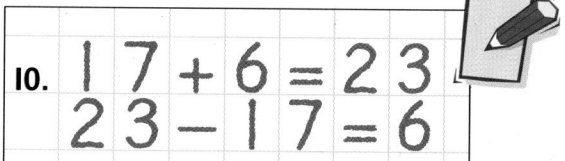

10. $17 + 6 = 23$
$23 - 17 = 6$

10
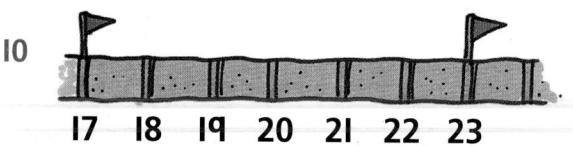

17　18　19　20　21　22　23

11
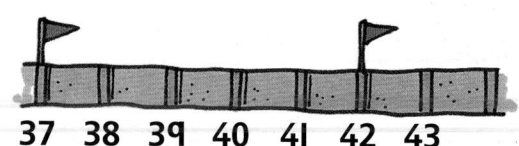

37　38　39　40　41　42　43

12
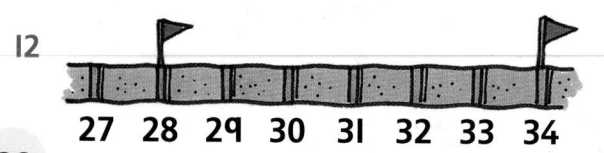

27　28　29　30　31　32　33　34

13
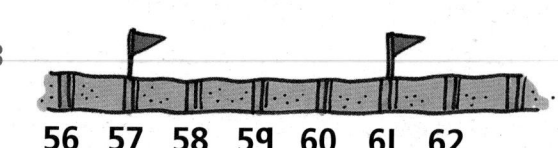

56　57　58　59　60　61　62

Difference

How much more money to buy each card?

1. **23p**

18p

1. $18p + 5p = 23p$
   $23p - 18p = 5p$

2. **25p**

17p

3. **32p**

25p

4. **44p**

37p

5. **34p**

28p

6. **52p**

45p

7. **71p**

66p

8.

Nellie is **37** today.

Nellie's brother Ned is **42**.

Write the difference in age.

9.
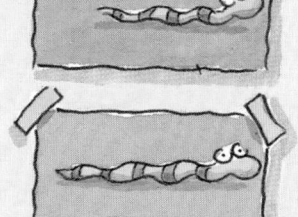
Sammy Snake was **29 cm** long.

Now he is **34 cm** long.

Write the difference in length.

Problems

10.
 POST

The van travelled **48 miles**.

This milk float travelled **53 miles**.

Write the difference in miles.

11.

The blue balloon went up **57 metres**.

The yellow balloon went up **65 metres**.

Write the difference in height.

37

Difference

Find the difference between the 2 card numbers.

1.

53 48

I. $53 - 48 = 5$

2.

42 39

3.

33 28

4.

21 16

5.

52 47

6.

22 18

7.

31 25

8.

45 54

9.

61 58

10.

65 72

 **Explore**

Complete the table.

Look carefully at the units.

Study the answers.

2<u>4</u>	− 19	= 5
3<u>6</u>	− 29	=
7<u>2</u>	− 69	=
4<u>4</u>	− 39	=
5<u>3</u>	− 49	=
1<u>2</u>	− 9	=
6<u>5</u>	− 59	=
8<u>7</u>	− 79	=

Can you see a pattern?

℮ Make up some similar subtractions of your own.

Adding three numbers

1

Write the total score.

15   35   12

1. $15 + 35 = 50$
   $50 + 12 = 62$

2
25   25   13

3
15   15   14

Look for multiples of 5.

4
5   75   18

5
15   65   12

6
35   35   21

7
45   15   16

8
55   25   12

9
65   15   13

10
75   14   25

11
28   5   65

12
45   17   15

Choose 3 targets and write the score.

Repeat 5 times.

25   15   35   45   12

e   Write the scores in order, from smallest to largest.

Adding three numbers

Write the total score on each screen.

I.
$$15 + 25 = 40$$
$$40 + 23 = 63$$

2

5 55 24

3

35 25 33

4

65 5 22

5

45 25 23

6

15 35 42

7

5 25 61

8

31 25 35

9

25 21 55

10

32 45 25

 **Explore**

The 3 cards total 100.

The ★ card is a
2-digit number.

Write down some possible
numbers for the star and heart cards.

  45

40

Adding three numbers

Write the total score.

I. $35 + 25 + 12 = 72$

1

12	25	14
45	17	35

2

12	25	14
45	17	35

3

12	25	14
45	17	35

4

12	25	14
45	17	35

5

12	25	14
45	17	35

6

12	25	14
45	17	35

℮ Write some other possible scores. What are the lowest and highest possible scores?

Problems

7 Alice Anteater eats **25** ants. Later she eats **35** more. She eats **24** before bedtime. How many in total?

8 Arni Armadillo eats **55** ants. Later he eats **15** more. He eats **28** before bedtime. How many in total?

9 Axel Aardvark eats **65** ants. Later he eats **25** more. He eats **18** before bedtime. How many in total?

Adding to 3-digit numbers

Copy and complete the additions.

1

2 3 2

232 + 8 =

I. $232 + 8 = 240$

2

1 4 3

143 + 7 =

3

1 5 4

154 + 6 =

4

3 7 8

378 + 2 =

5

1 8 6

186 + 4 =

6

2 7 5

275 + 5 =

7

1 3 1

131 + 9 =

8

2 5 2

252 + 8 =

9

1 9 6

196 + 4 =

10

3 3 3

333 + 7 =

Add to each score.

Write each total score.

11

146

Add 20

II. $146 + 20 = 166$

12

255

Add 30

13

374

Add 20

14

417

Add 40

15

525

Add 50

Adding to 3-digit numbers

Add 6p to each sack of coins.

1. $126p + 6p = 132p$

1 126p

2 164p

3 128p

4 137p

5 155p

6 228p

7 177p

8 293p

e Add 16p to each sack.

Problems

9
Fred saves 1p coins.

155p

He is given **20p** more.

How much has he now?

10
Bill collects football stickers.

146 stickers

He is given **9** more.

How many has he now?

11
Ana collects stamps.

246 stamps

She collects **8** more.

How many has she now?

Adding to 3-digit numbers

Copy and complete.

1 124 + 22 =

2 136 + 23 =

3 245 + 25 =

4 150 + 36 =

5 346 + 31 =

6 444 + 44 =

7 285 + 15 =

8 625 + 25 =

9 348 + 51 =

10 255 + 31 =

11 152 + 34 =

12 176 + 13 =

13 224 + 24 =

Explore

Use the cards shown.

Arrange them to make an addition like this.

1 2 3 4

1 2 3 + 4 ⟶ 123 + 4 =

Write the answer.

How many different additions can you make?

ℯ Write the smallest and largest totals.

Adding multiples of one hundred

Write how many people in total.

1. $2 4 1 + 3 0 0 = 5 4 1$

1 | 241 | 300

2 | 366 | 400

3 | 414 | 500

4 | 179 | 500

5 | 196 | 200

6 | 344 | 600

7 | 374 | 400

8 | 299 | 300

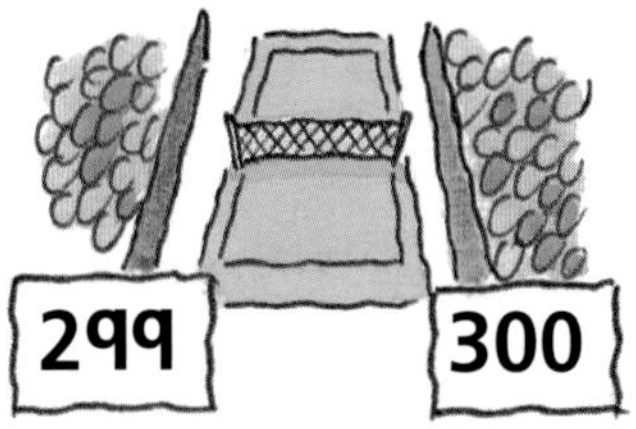

9 | 458 | 500

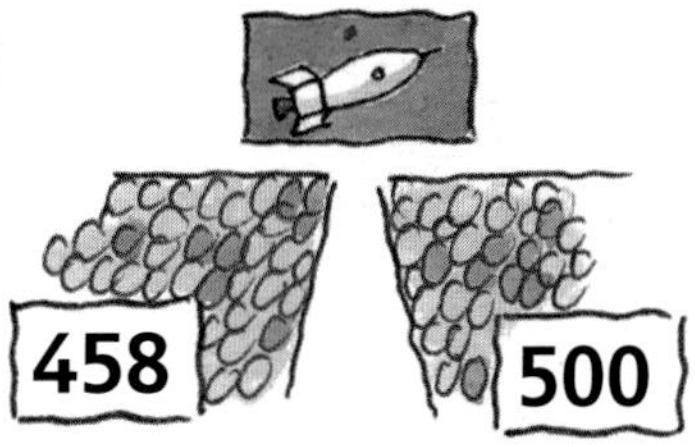

10 | 213 | 200

11 | 421 | 400

12 | 444 | 500

Add 250 to each star number.

13. $3 4 2 + 2 5 0 = 5 9 2$

13 342

14 431

15 125

16 616

17 249

18 750

Adding two 3-digit numbers

Write the total.

I. $423 + 200 + 6 = 629$

1 | 4 2 3 | 2 0 6
423 + 200 + 6 = ◯

2 | 5 5 5 | 1 0 4
555 + 100 + 4 = ◯

3 | 3 2 4 | 2 0 2
324 + 200 + 2 = ◯

4 | 4 1 2 | 3 0 7
412 + 300 + 7 = ◯

5 | 8 2 1 | 1 0 8
821 + ◯ + ◯ = ◯

6 | 4 5 7 | 4 0 1
457 + ◯ + ◯ = ◯

7 | 2 2 1 | 5 0 8
221 + ◯ + ◯ = ◯

8 | 6 4 4 | 3 0 3
644 + ◯ + ◯ = ◯

Choose 2 items.

q. $£3·42 + £2·10 = £5·52$

Write the total cost.

Repeat 10 times.

£2·12

£4·24

£3·42

£1·10

£2·10

£3·20

Subtracting 3-digit numbers

Subtract the penalty fine from each score.

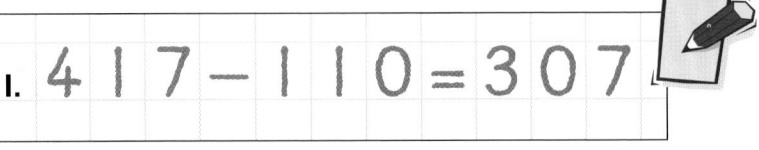

1. $417 - 110 = 307$

1

SCORE 417
PENALTY 110

2
SCORE 667
PENALTY 220

3
SCORE 426
PENALTY 210

4

SCORE 576
PENALTY 240

5
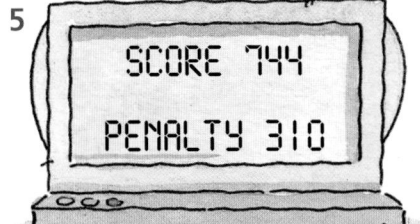
SCORE 744
PENALTY 310

6

SCORE 352
PENALTY 120

7

SCORE 474
PENALTY 150

8

SCORE 691
PENALTY 230

9

SCORE 535
PENALTY 320

The wind changes direction.

Each balloon drops in height.

Write the new height.

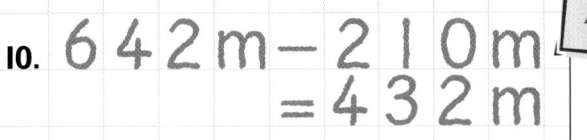

10. $642\,m - 210\,m = 432\,m$

10
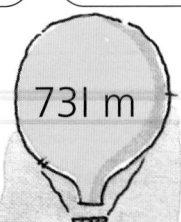
642 m
drops 210 m

11
731 m
drops 320 m

12
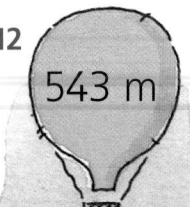
543 m
drops 130 m

13
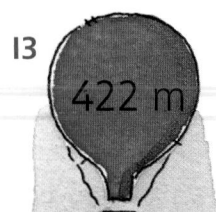
422 m
drops 210 m

14

311 m
drops 110 m

Odd and even

Copy and write the missing numbers.

1. 2, 4, 6, 8, 10, 12

1 | 2 | 4 | | 8 | | |

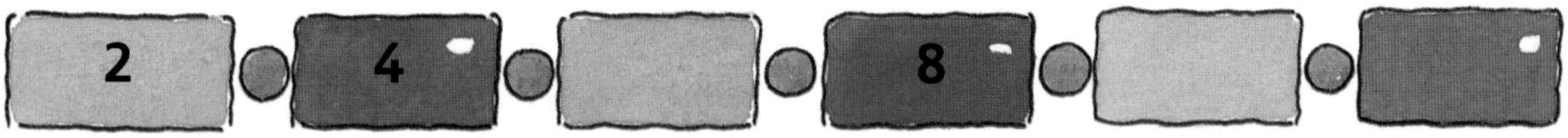

2 | 1 | 3 | 5 | | | 13 | | | | 23 |

3 | 34 | 36 | 38 | | | | | | | | 56 |

4 | 61 | 63 | 65 | | | | | | | | 83 |

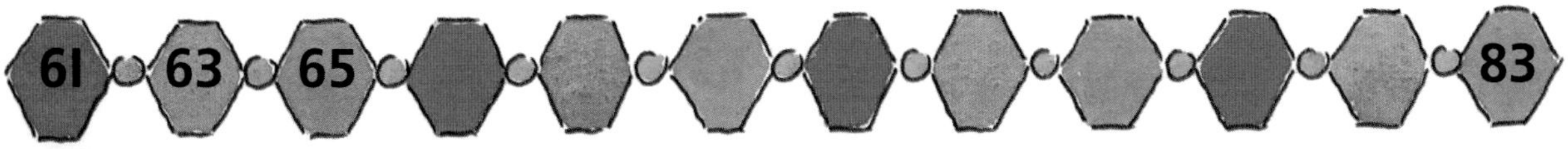

5 | 70 | 72 | 74 | | | | | | | | 92 |

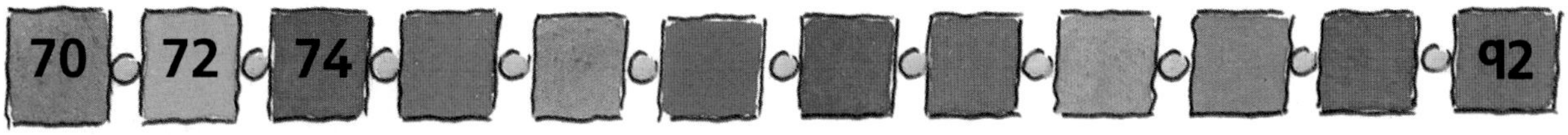

6 | 43 | 45 | | | | | | | | 63 | 65 |

Write odd or even next to each number.

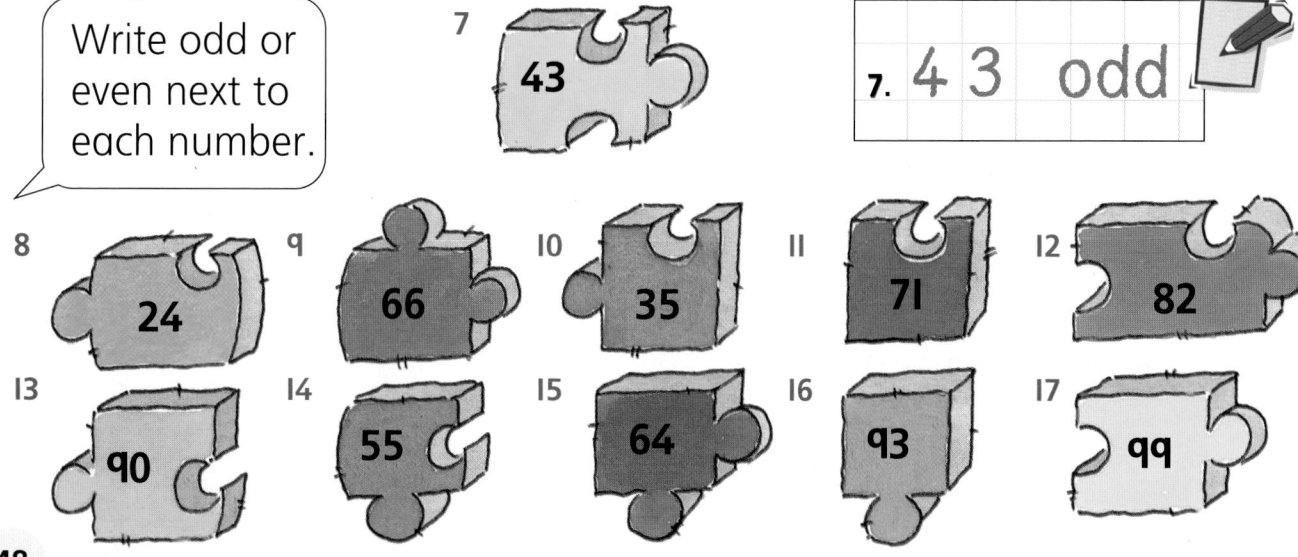

7 43

7. 43 odd

8 24

9 66

10 35

11 71

12 82

13 90

14 55

15 64

16 93

17 99

48

Odd and even

Each child needs a unit card to make their number match their cap.

Write their number.

I

I. even 4 4

Place-value cards

2

3

4

5

6

7

8

9

e Write the next odd or even number for each answer.

 **Explore**

How many odd numbers are there with an even tens digit, between 1 and 100?

Remember to use a system.

Odd and even

Write odd or even next to each price.

I. 7 4 p even

1. 74p

2. 45p

3. 51p

4. 66p

5. 92p

6. 54p

7. 63p

8. 59p

9. 78p

10. 28p

11. 33p

12. 97p

13. 81p

Explore

Sam and Sanjit have **2I stickers** altogether.

Sam has an odd number.
Sanjit has an even number.

Write the different amounts they could each have.

Remember to use a system.

Multiplying by 10 and 100

Write a multiplication for each set.

1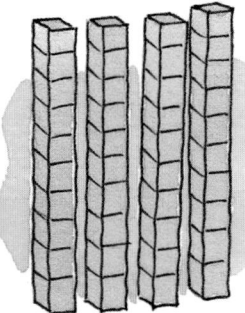

I. $4 \times 10 = 40$

2

3

4

5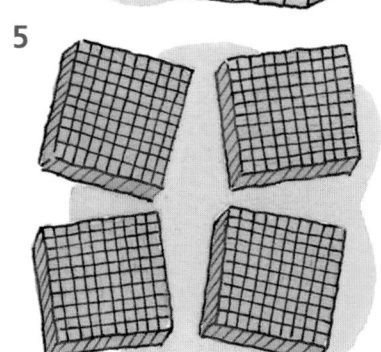

6

7

Write how many pennies match each amount.

8

8. $3 \times 100 = 300p$

9

10

11

12

13

14

Multiplying by I0 and I00

Write the height of each object in centimetres.

I. $6 \times 100 = 600$ cm

I metre = I00 centimetres

1 6 m

2 9 m

3 2 m

4 7 m

5 3 m

6 5 m

7 8 m

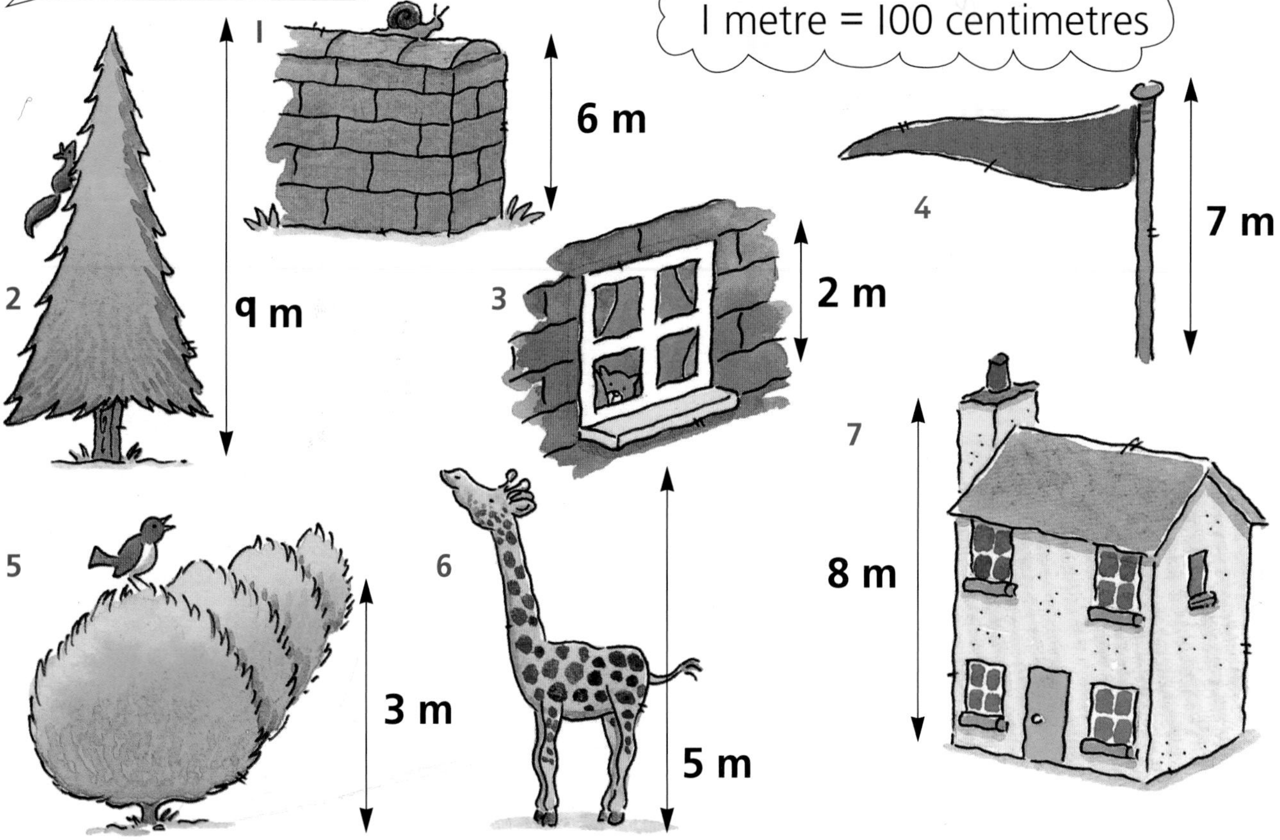

Copy and complete.

8 $4 \times 10 =$

8. $4 \times 10 = 40$

9 $2 \times 10 =$

10 $3 \times 100 =$

11 $7 \times 10 =$

12 $5 \times 100 =$

13 $9 \times 10 =$

14 $1 \times 100 =$

15 $6 \times 10 =$

16 $7 \times 100 =$

17 $1 \times 10 =$

18 $12 \times 10 =$

19 $4 \times 100 =$

20 $15 \times 10 =$

Multiplying by 10 and 100

There are 10 crayons in each box.

Write how many crayons in each set.

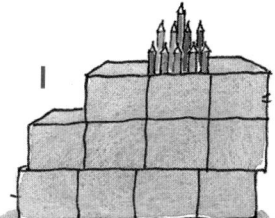

1.

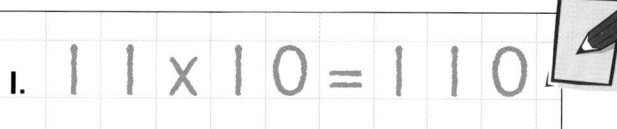

I. $11 \times 10 = 110$

2

3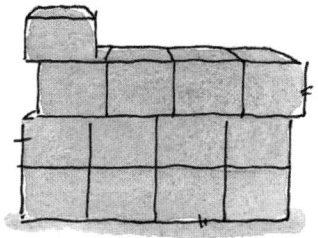

4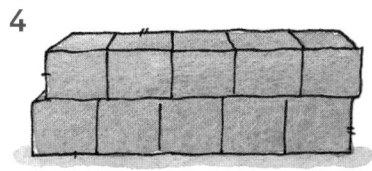

5

6

7

8

q

10

Karen has **10** maths questions every day at school.

How many questions after I week?

Problems

11

Keri runs around the track **12** times.

It is **100 m** all the way round.

How far has she run?

12

There are **10** children round each table at lunchtime.

There are **14** tables.

How many children?

Multiplying

Kim   Toni   Sanjit   Drew   Tanya

> These children save their pocket money every week.

> Write how much after 2 weeks.

Kim
2 x 3 0 p = 6 0 p

> Write how much after 3 weeks.

Kim
3 x 3 0 p = 9 0 p

> Write how much after 4 weeks.

Kim
4 x 3 0 p = 1 2 0 p
= £ 1·2 0

> Find the total length round each shape.

I. 4 x 2 0 cm = 8 0 cm

1	2	3	4	5
20 cm		←40 cm→	←30 cm→	20 cm
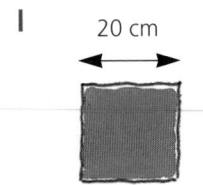	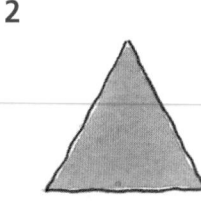	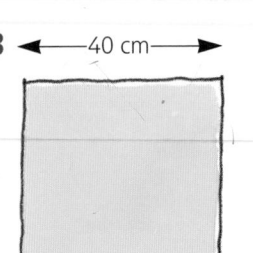	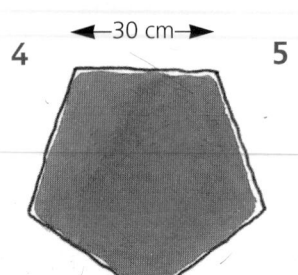	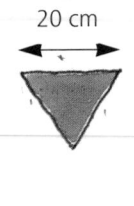
	←30 cm→			

Multiplying

 20p

 30p

 40p

 50p

Write the cost to post each letter.

1

1. $2 \times 20p = 40p$

2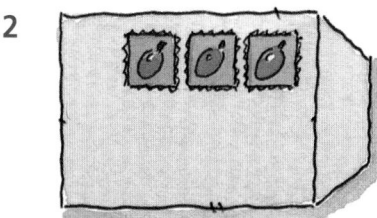

3

4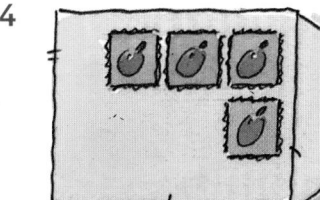

5

6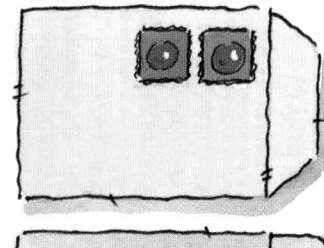

7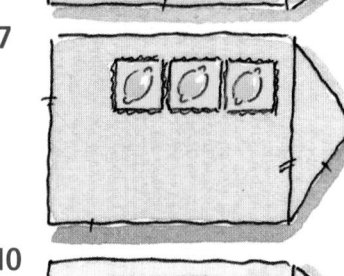

8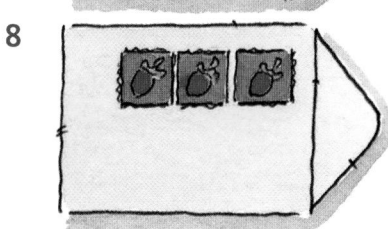

9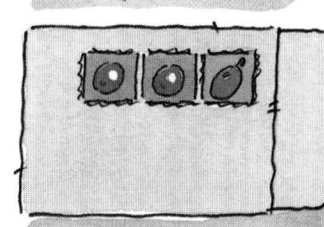

10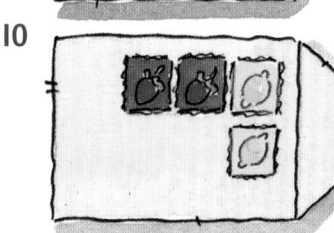

Explore

Use the cards shown.

Use $\boxed{0}$ each time and two other cards.

Make different multiplications like this: $\boxed{2} \times \boxed{5}\boxed{0} =$

Write each multiplication and find the total.

How many different totals can you find? How many have the same total? What do you notice?

Multiplying

Write the score on each screen.

I. $2 \times 20 = 40$

Copy and complete.

7 $3 \times 30 = $

7. $3 \times 30 = 90$

8 $4 \times$ ▓ $= 120$ 9 $5 \times$ ▓ $= 100$ 10 ▓ $\times 30 = 60$

11 ▓ $\times 50 = 200$ 12 $4 \times$ ▓ $= 160$ 13 $2 \times$ ▓ $= 100$

Problems

14 There are **10** chocolate buttons in each pack shown.

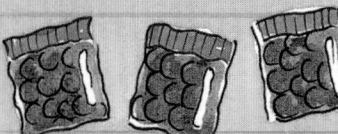

How many buttons in total?

15 April, June, September and November have **30 days**.

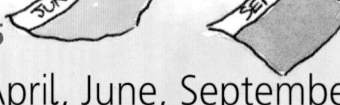

How many days is that altogether?

16

The wall is **40 m** away. Jo runs there and back.

How far has she run?

Fours

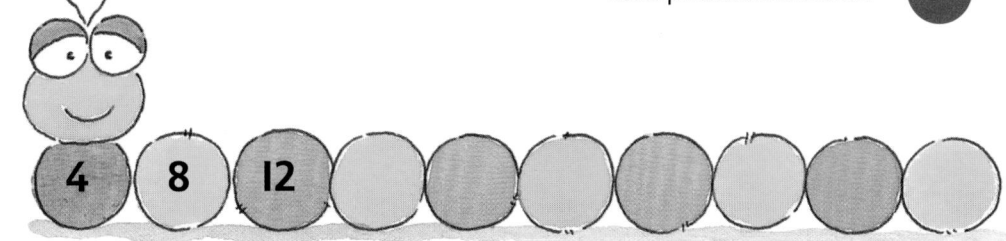

Copy and complete.

| 4 | 8 | 12 | | | | | | | |

How many legs in each group of dogs?

 1

1. $2 \times 4 = 8$

 2

 3

 4

 5

 6

 7

 8

 9

Copy and complete.

10. $1 \times 4 = 4$

10 $1 \times 4 =$ 11 $6 \times 4 =$ 12 $9 \times 4 =$ 13 $2 \times 4 =$ 14 $10 \times 4 =$

15 $7 \times 4 =$ 16 $4 \times 4 =$ 17 $3 \times 4 =$ 18 $5 \times 4 =$ 19 $8 \times 4 =$

e Write these facts in order.

Fours

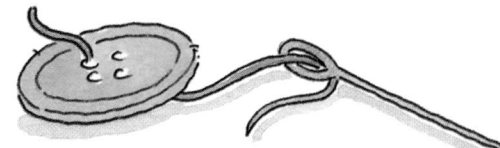

> Each button has 4 holes.
> Write a multiplication and division for each set.

1

I. $3 \times 4 = 12$
$12 \div 4 = 3$

2

3

4

5

6

7

8

9

10

> Copy and complete.

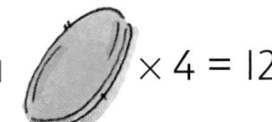

11 $\bigcirc \times 4 = 12$

II. $3 \times 4 = 12$

12 $\bigcirc \times 4 = 8$

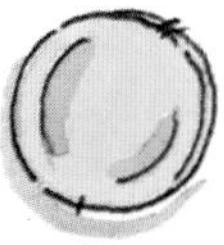

13 $\heartsuit \times 4 = 24$

14 $\bigcirc \div 4 = 5$

15 $\triangle \div 4 = 4$

16 $\star \times 4 = 32$

17 $\triangle \div 4 = 1$

18 $\bigcirc \times 4 = 28$

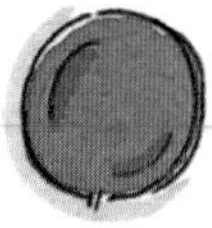

19 $\bigcirc \times 4 = 36$

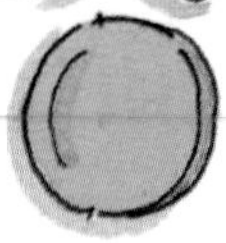

20 $\bigcirc \div 4 = 10$

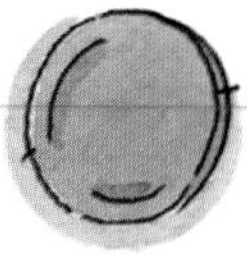

Fours

Write how long the sides of each **square** field are.

1

28 m in total

I. $28m \div 4 = 7m$

2

16 m in total

3

12 m in total

4

32 m in total

5

4 m in total

6

20 m in total

7

24 m in total

Explore

Copy the grid and colour the numbers in the ×4 table.

Draw a grid of your own with 6 columns and 6 rows.

Colour the numbers in the ×4 table.

Look for patterns.

Try different grids.

1	2	3	4	5
6	7	8	9	10
11	12	13	14	15
16	17	18	19	20
21	22	23	24	25
26	27	28	29	30
31	32	33	34	35
36	37	38	39	40

Dividing

Write a division for each set.

I

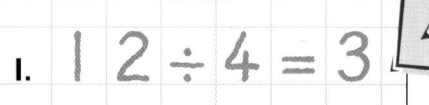
I. $12 \div 4 = 3$

2

3

4
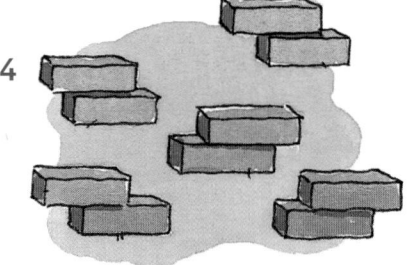

5

6

7

8

9

10

e Write a multiplication for each set.

Copy and complete.

Counters

II. $14 \div 2 = 7$

11 $14 \div 2 =$

12 $8 \div 4 =$

13 $10 \div 5 =$

14 $16 \div 2 =$

15 $20 \div 10 =$

16 $15 \div 5 =$

17 $21 \div 3 =$

18 $16 \div 4 =$

19 $30 \div 10 =$

Remainders

There are 14 children in the Abacus Infant football club.

The children are put into practice teams.

Write the division and any remainders.

1   teams of 3

1. $14 \div 3 = 4 \text{ r } 2$

2   teams of 4

3   teams of 2

4   teams of 5

There are 18 children in the Junior club.

5   teams of 5

6   teams of 2

7   teams of 4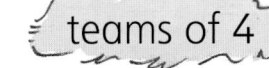

8   teams of 3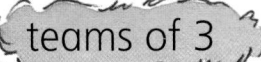

9   teams of 6

10   teams of 8

Explore

 4   8   12   16   20   24    28   32   36   40

Take each number from the ×4 table in turn.

Match it with counters and arrange them in groups of 3.

Write the remainder each time.

Describe the pattern.

Explore dividing numbers in the ×5 or ×6 tables by 3.

Counters.

Remainders

Each car can take 4 children.

How many cars to take each group?

> 1. $22 \div 4 = 5 \text{ r } 2$
> → 6 cars

1 22 children

2 24 children

3 18 children

4 20 children

5 17 children

6 13 children

Football match tickets are £3.00.

How many can each person buy?

7 £17.00

8 £21.00

9 £31.00

Problems

10
There are **27** children.

How many teams of **5**?

11
Kelly has **27p**.
Stickers are **4p** each.

How many can she buy?

How much to buy **9** stickers?

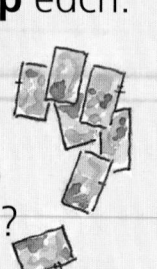

12
Gary has **17** biscuits.

He eats **3** and gives **3** each to **4** friends.

How many are left?

Matching fractions

Write the fraction
of each circle
that is pink.

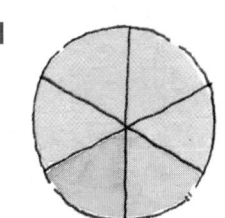

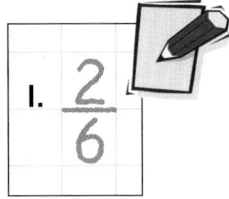

I. $\dfrac{2}{6}$

2 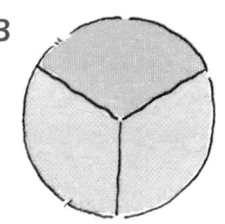 3 4 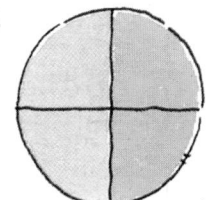 5 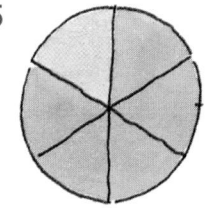 6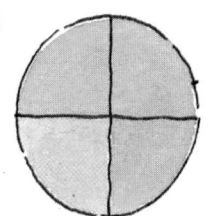

7 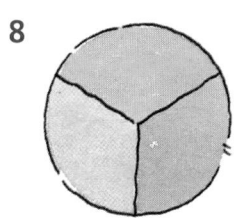 8 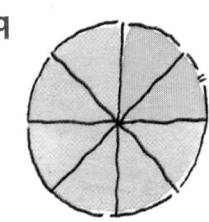 9 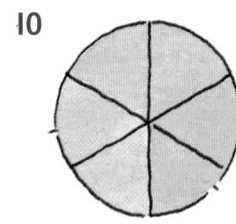 10

e Write the pairs that show matching fractions.

Draw 2 × 4 grids.

Copy and write the
matching fraction for
the coloured squares.

11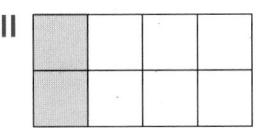

II.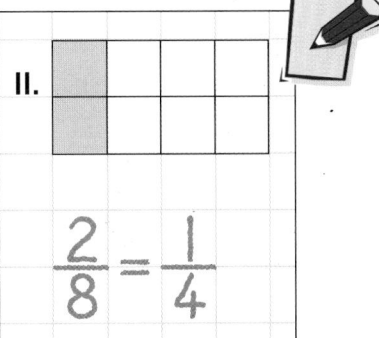

$\dfrac{2}{8} = \dfrac{1}{4}$

12 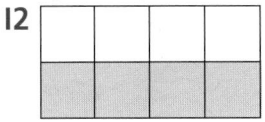 13 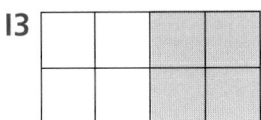 14 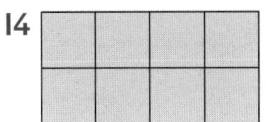 15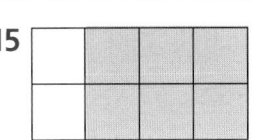

16 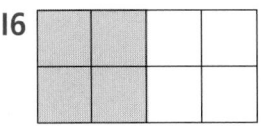 17 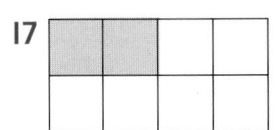 18 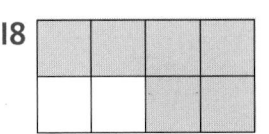 19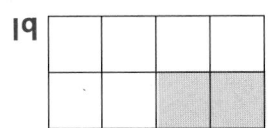

Matching fractions

Write pairs
of fractions
that match.

1. $\dfrac{1}{2} = \dfrac{2}{4}$

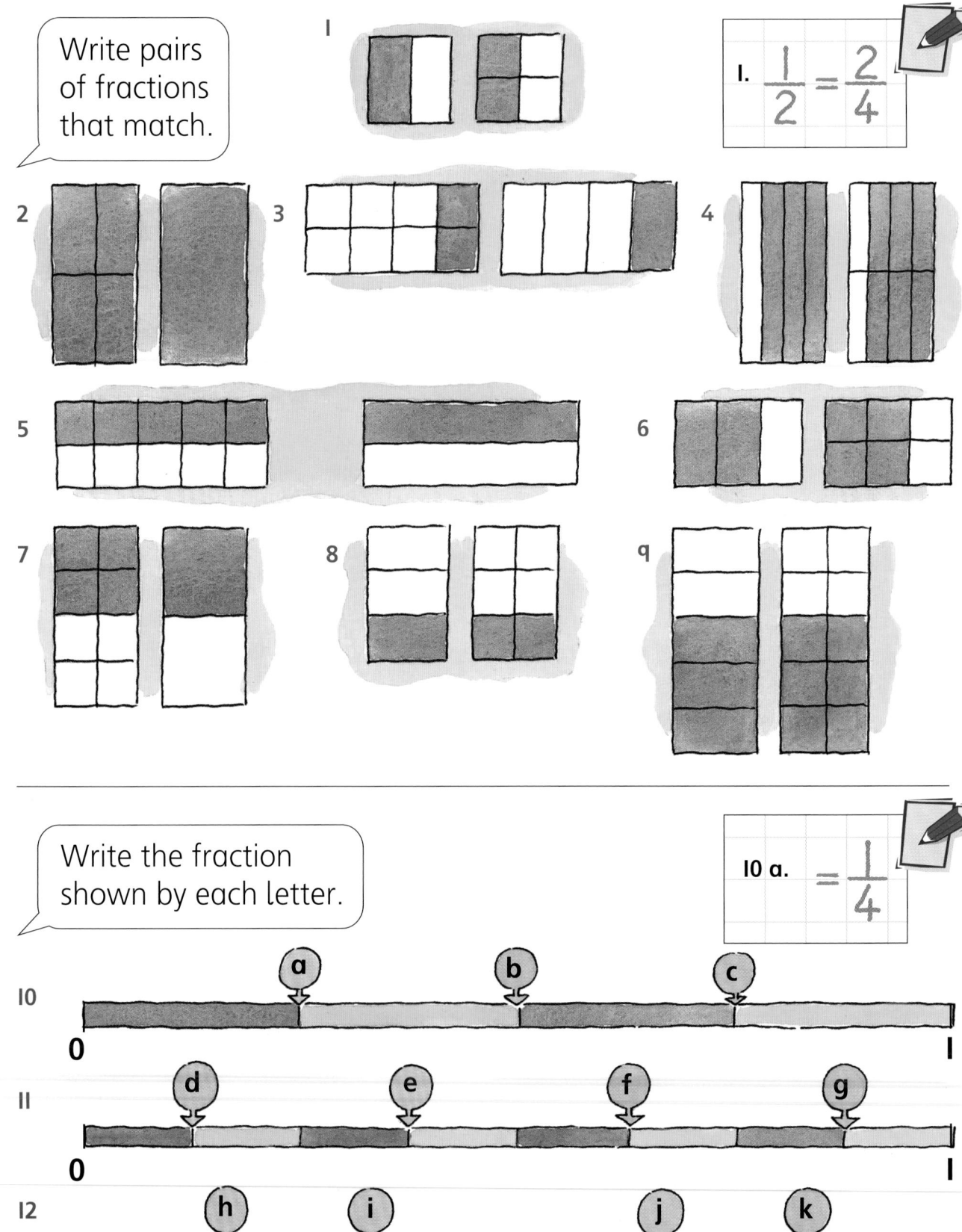

Write the fraction
shown by each letter.

10 a. $= \dfrac{1}{4}$

Matching fractions

Copy and complete.

I. $\dfrac{1}{4} = \dfrac{2}{8}$

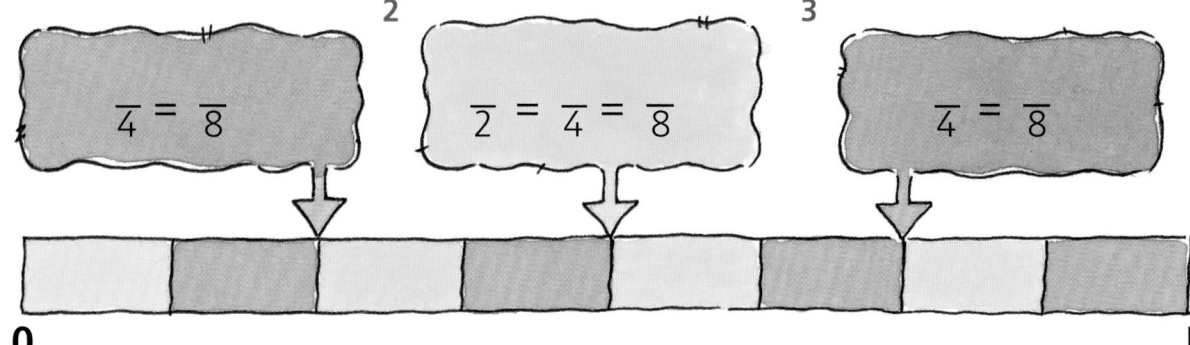

1. $\dfrac{\ }{4} = \dfrac{\ }{8}$

2. $\dfrac{\ }{2} = \dfrac{\ }{4} = \dfrac{\ }{8}$

3. $\dfrac{\ }{4} = \dfrac{\ }{8}$

0   1

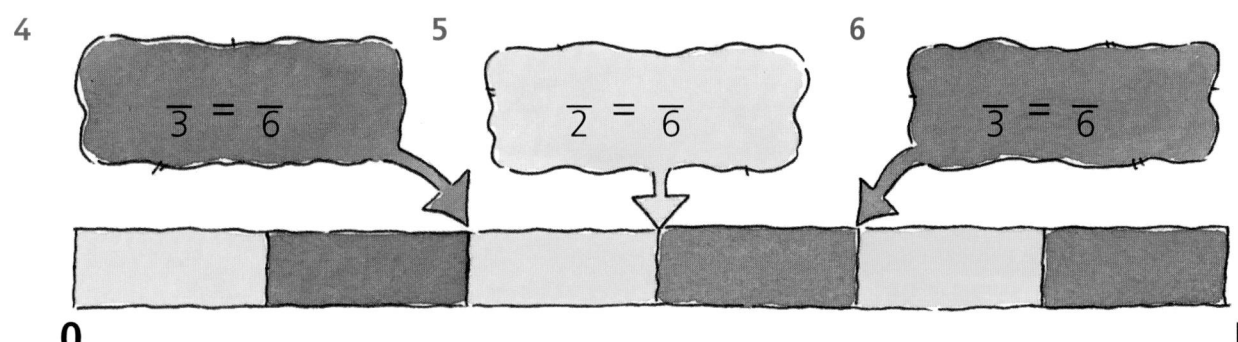

4. $\dfrac{\ }{3} = \dfrac{\ }{6}$

5. $\dfrac{\ }{2} = \dfrac{\ }{6}$

6. $\dfrac{\ }{3} = \dfrac{\ }{6}$

0   1

Write the matching pairs.

7. $\dfrac{1}{2} = \dfrac{4}{8}$

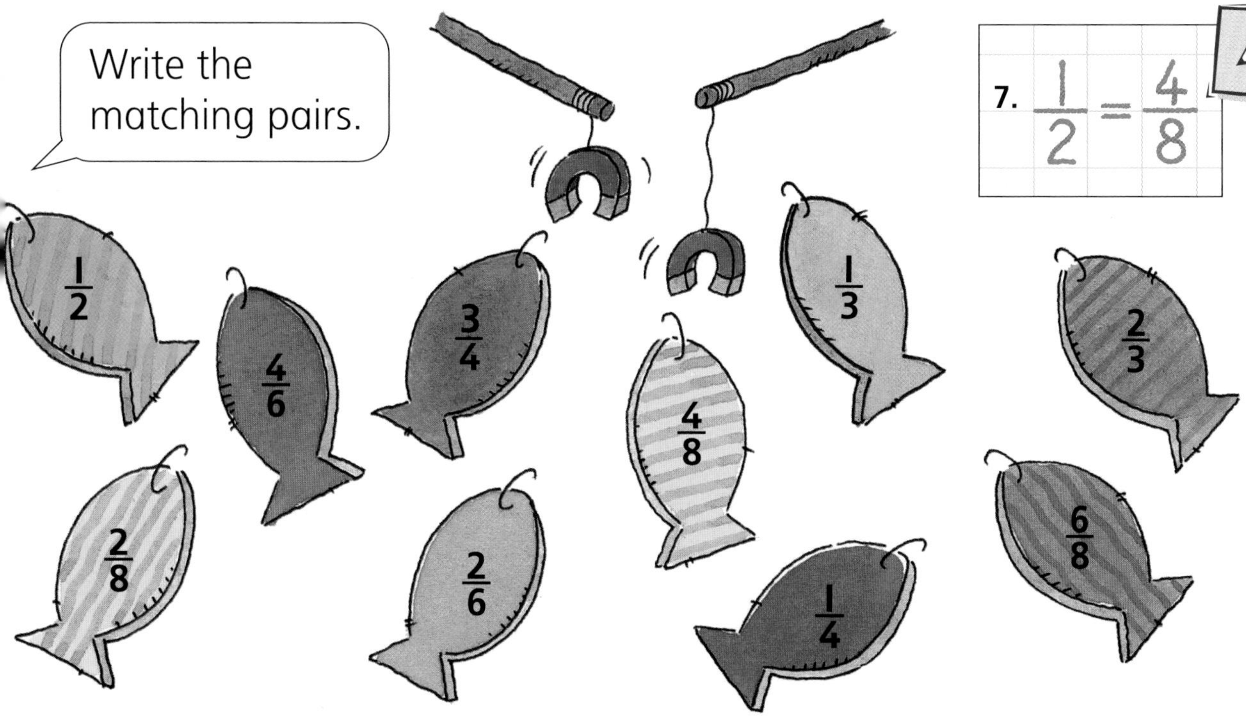

$\dfrac{1}{2}$ $\dfrac{4}{6}$ $\dfrac{3}{4}$ $\dfrac{4}{8}$ $\dfrac{1}{3}$ $\dfrac{2}{3}$

$\dfrac{2}{8}$ $\dfrac{2}{6}$ $\dfrac{1}{4}$ $\dfrac{6}{8}$

65

Adding two 3-digit numbers

> Write the total for each pair.

I. $243 + 100 = 343$
$343 + 9 = 352$

1
243   109

2
464   207

3
555   306

4
208   352

5
104   277

6
388   404

7
626   205

8
108   724

9
357   303

10
307   485

11
538   203

12
612   212

> Copy and complete.

13   $436 + 160$

14   $545 + 140$

15   $438 + 240$

16   $664 + 320$

17   $527 + 250$

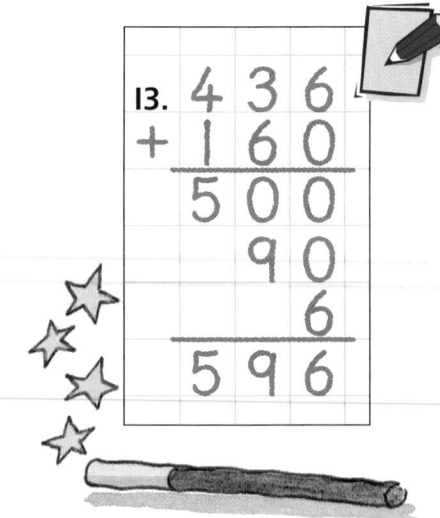

13.
```
  4 3 6
+ 1 6 0
  5 0 0
    9 0
      6
  5 9 6
```

Adding two 3-digit numbers

Write the total score for each juggler.

I.
```
   4 2 7
 + 2 3 1
   6 0 0
     5 0
       8
   6 5 8
```

I 427   231

2 346   323

3 251   317

4 511   488

5 104   125

6 416   382

7 383   415

8 220   219

q 191   308

10 281   311

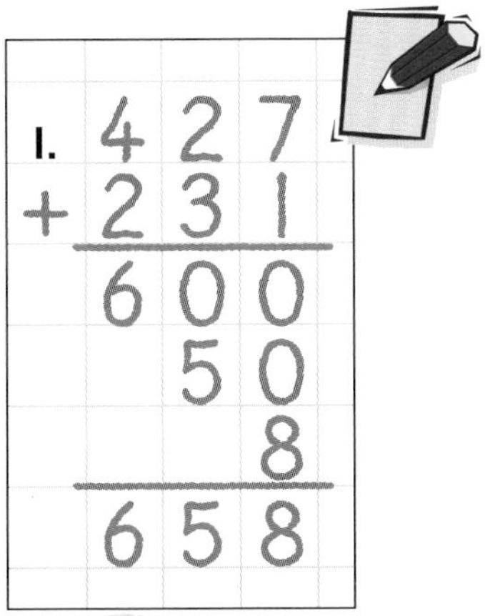

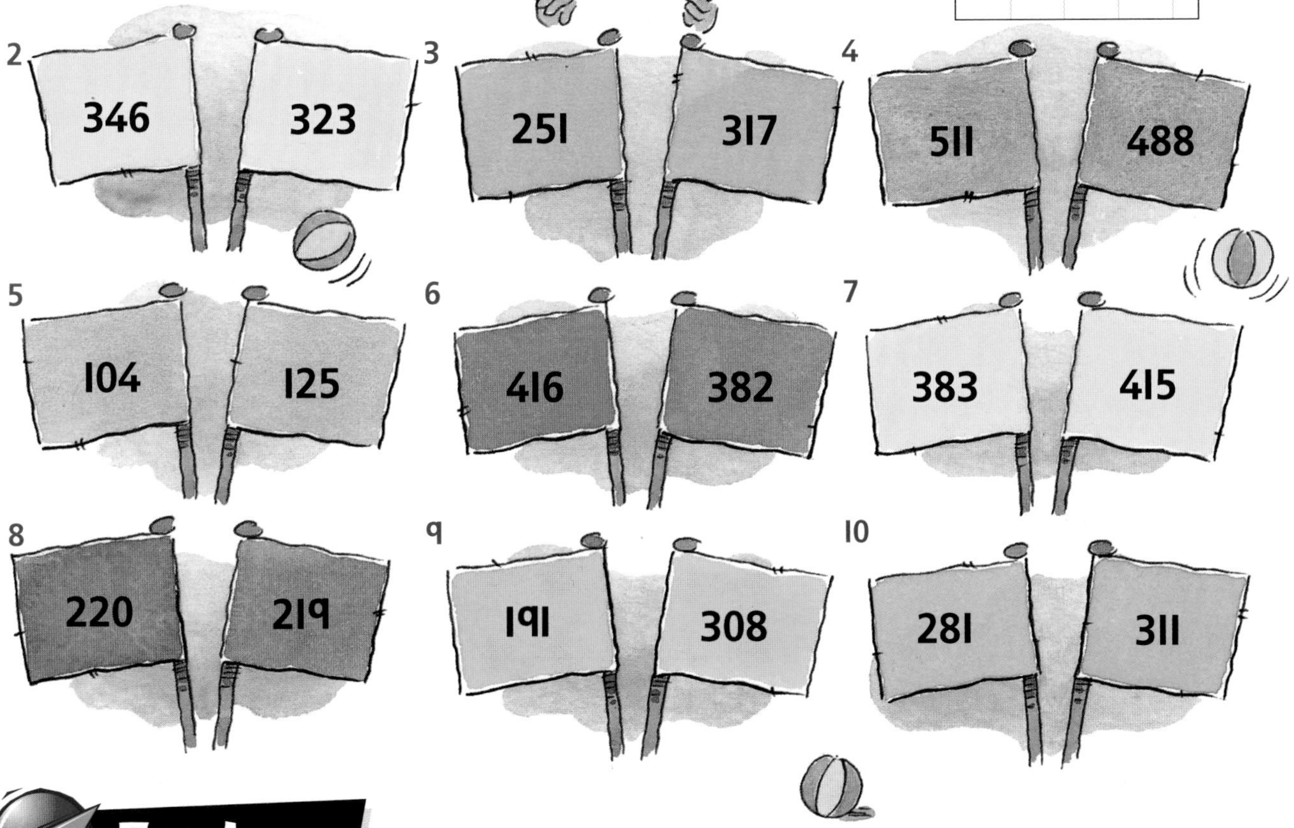

Explore

Use the cards shown.

Make a 3-digit number.

Write it down.

Reverse the cards and write the new number.

Add the two numbers.

Repeat 8 times. What do you notice?

0 4 3 2 I

421 + 124 =

Adding two 3-digit numbers

I. £512 + £430 = £942

Choose 2 safes.

Write the total amount.

Repeat 10 times.

£241

£353

£342

£512

£425

£134

£430

e Add £255 to each safe.

Problems

11 Sally has **428** jigsaw pieces.

Sam has **216** pieces.

They help each other to finish the jigsaw.

How many pieces altogether?

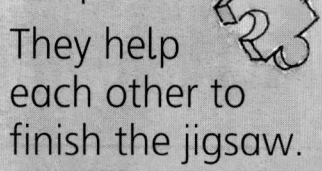

12 Jem swims **134 m** on Monday.

He swims **219 m** on Friday.

How many metres altogether?

13 Aki's gran has knitted **532** rows of a scarf.

Aki knits **144** rows.

How many rows altogether?

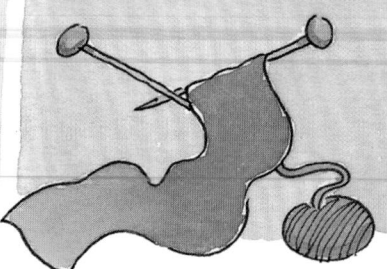

Adding 3-digit numbers

Add the units first.

H	T	U
3	7	6
+ 2	5	6

Write the ten in the tens column.

H	T	U
3	7	6
+ 2	5	6
		2
1		

Add the tens.

H	T	U
3	7	6
+ 2	5	6
		2
	1	

Add the hundreds and read the total.

H	T	U
3	7	6
+ 2	5	6
6	3	2
	1	1

Copy and complete.

1

H	T	U
1	7	5
+ 2	3	7

2

H	T	U
3	2	5
+ 4	8	6

3

H	T	U
3	6	4
+ 1	6	8

4

H	T	U
2	9	1
+ 6	1	9

5

H	T	U
7	2	4
+ 1	5	6

6

H	T	U
4	4	4
+ 4	5	9

7

H	T	U
2	4	7
+ 6	4	7

8

H	T	U
1	9	5
+ 5	4	5

9

H	T	U
4	5	9
+ 1	5	8

10

H	T	U
2	6	7
+ 5	4	4

11

H	T	U
3	1	6
+ 4	9	9

12

H	T	U
2	3	9
+ 5	6	6

Adding 2-digit and 3-digit numbers

> Add 158 g of sugar to each mixture.

> Write the total weight.

1 127 g    Sugar

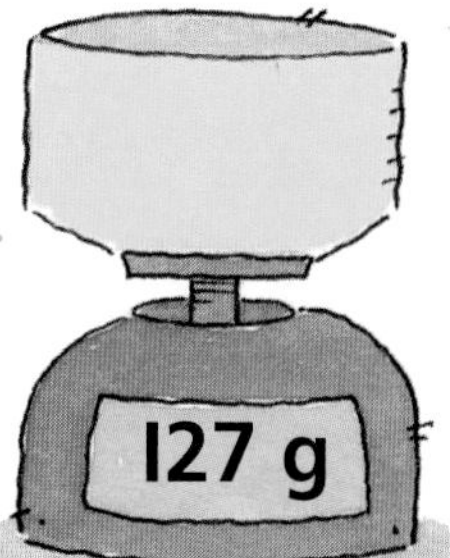

```
   H T U
1.
   1 2 7
 + 1 5 8
   2 8 5 g
     1
```

2 248 g

3 347 g

4 384 g

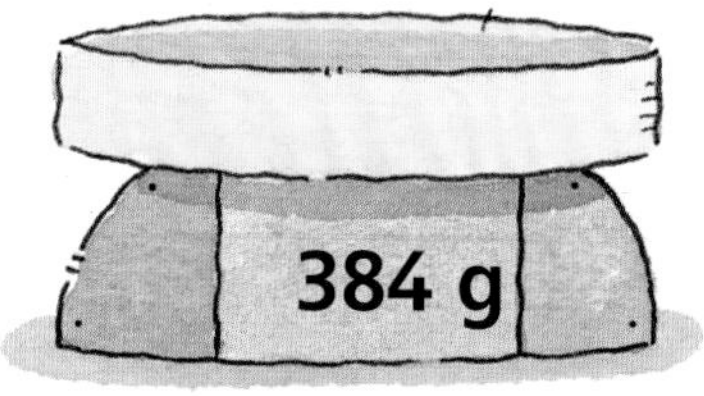

5 492 g

6 559 g

7 649 g

8 199 g

9 632 g

10 408 g

> Choose a pencil and a rubber each time.

> Write the total.

> Repeat 5 times.

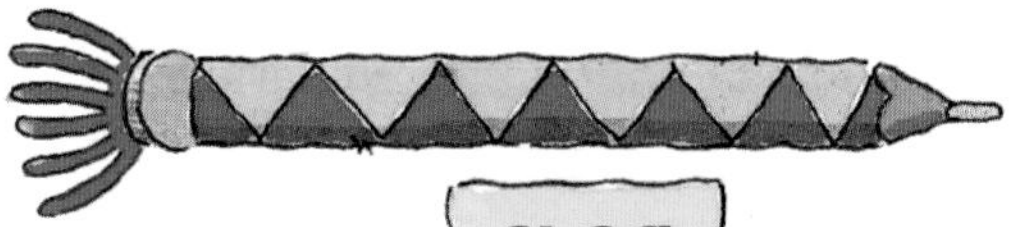

£1·29

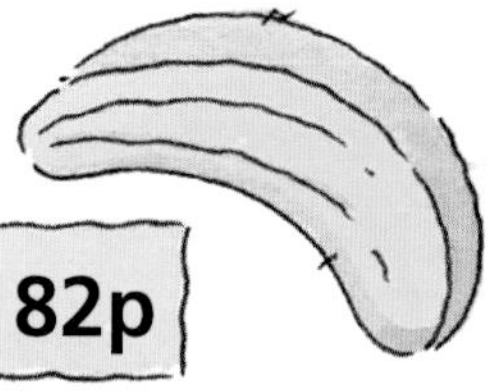

82p

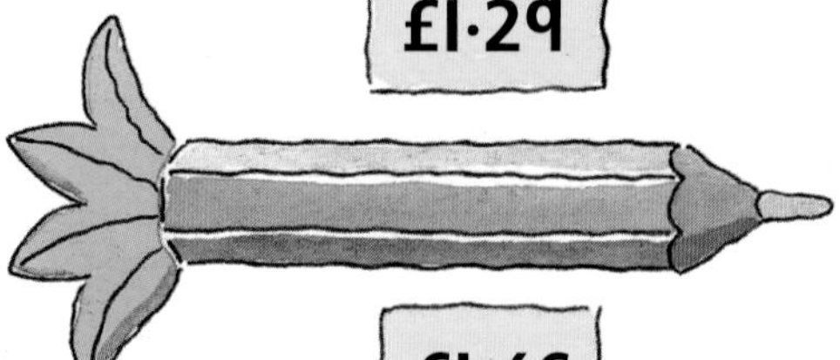

£1·46

77p

£1·55

69p

✏ What is the most money you would need?

Adding 3-digit numbers

Write the total weight.

1
468 g 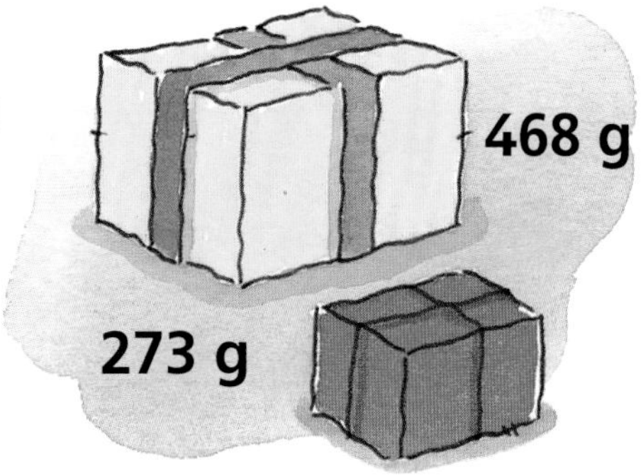
273 g

```
   4 6 8
 + 2 7 3
   7 4 1  g
   1 1
```

2
137 g 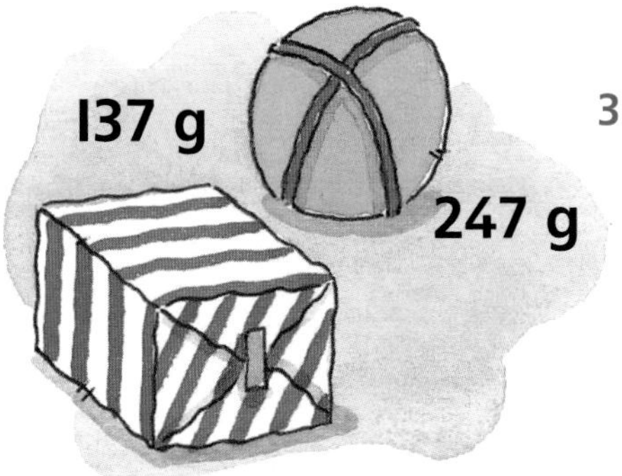
247 g

3
683 g 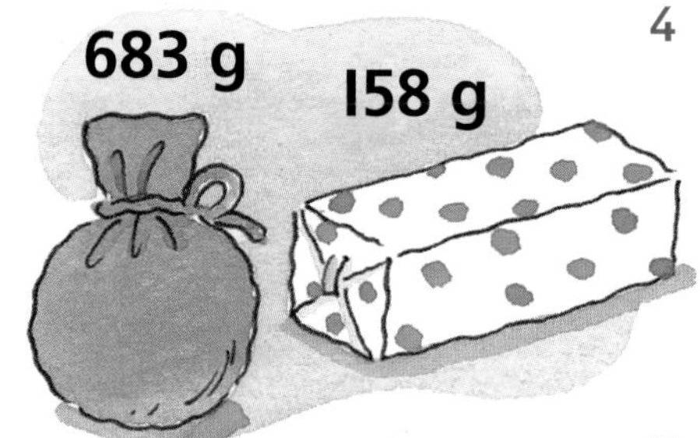
158 g

4
627 g 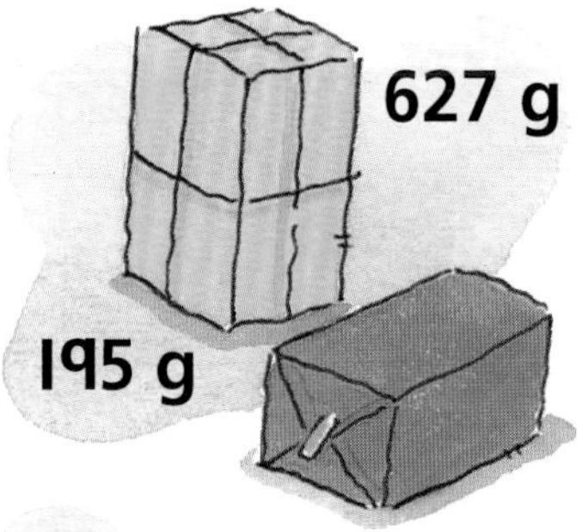
195 g

5
246 g 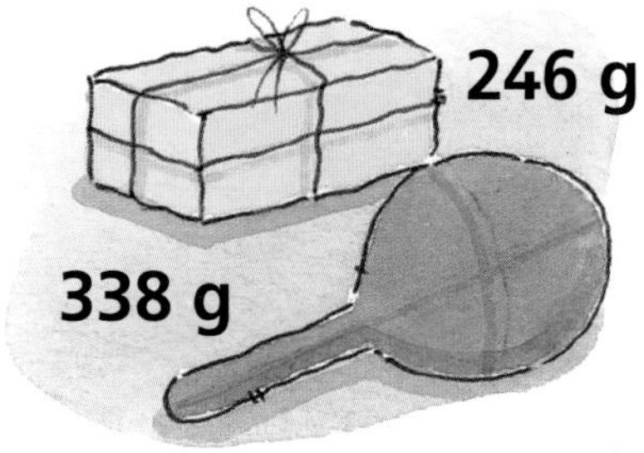
338 g

6
399 g 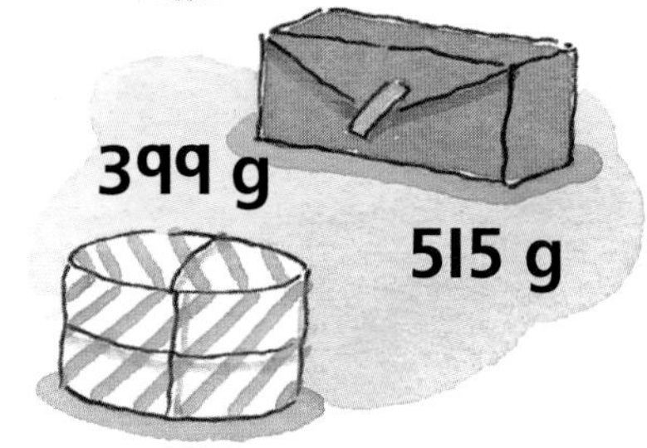
515 g

7
434 g 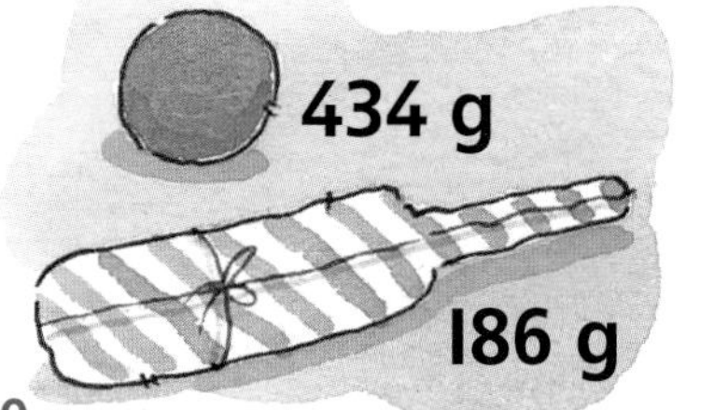
186 g

8
738 g
193 g

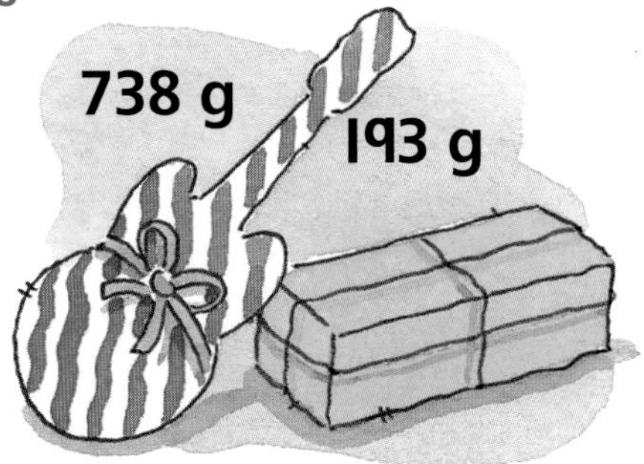

9
714 g
117 g

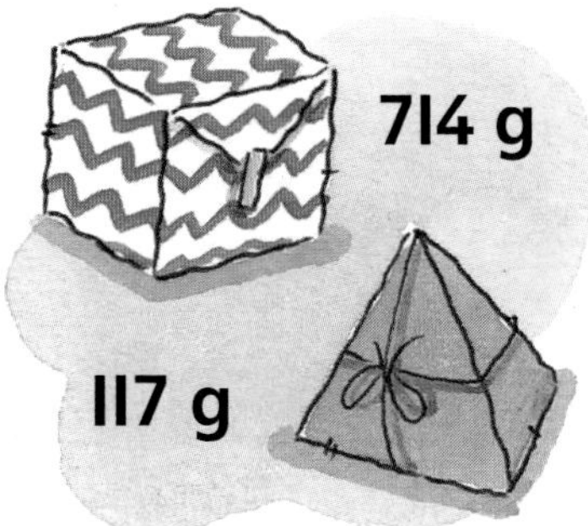

10
326 g
495 g

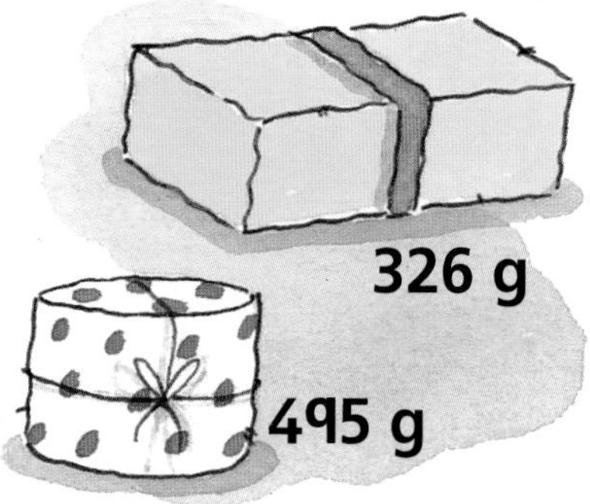

Explore

Use the cards shown.

Make a 3-digit number and a 2-digit number.

Add them together.

1 2 3 4 5

Try to find ways to make these answers: 168   483   339

Mixed problems

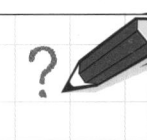

> Each player has 2 number cards.

> Use the information to decide which cards they have.

1. If you multiply the numbers your total is 16.
   If you add the numbers your total is 10.

2. If you add the numbers your total is 15.
   The difference between them is 1.

3. One number is half the other number.
   They add up to a total of 24.

4. Both numbers have the same 2 digits but they are not the same number.
   They add up to 33.

> Each of the children has some money in their purse.

> Use the information to decide how much.

Tara   Harpreet   Winston   Ben

5. Winston has 3 coins.
   The first is double the second.
   The second is double the third.

6. If Tara buys 3 toys at £1.70, 50p and 25p, she will still have 55p left in her purse.

7. Ben has two of every silver coin.

8. Harpreet can put two 20p stamps and three 30p stamps on her mum's parcel.
   She will have 5p left.